D0904833

# About the author

Colin Tudge is a biologist by education and a writer by trade with a lifelong passion for food and agriculture. In 2004 Colin coined the term 'enlightened agriculture', which gave rise to the Campaign for Real Farming. Together with his wife, Ruth (West) and the agricultural writer Graham Harvey, he co-founded the Oxford Real Farming Conference in 2010; and in 2012, he and Ruth launched FEA (Funding Enlightened Agriculture), a network to support enterprises that operate according to principles of enlightened agriculture. Now he has established the College for Real Farming and Food Culture to explore and promulgate all the underlying ideas.

The good news is that we can feed everyone, and can do that in healthy ways indefinitely into the future. But if we are to do this while minimising the disastrous impacts that accompany modern industrial farming, then we'll need to do things differently. With characteristic clarity Colin Tudge lays out why this is not primarily a technical issue but is more political and economic in nature.
***Tony Juniper former Executive Director of Friends of the Earth, author of* What Has Nature Ever Done for Us?**

Colin Tudge offers a welcome perspective at a time when so much of the dialogue about food and farming is framed against a backdrop of fear; a fear that there won't be enough to feed an ever-increasing world population and a fear of the apocalyptic consequences of agriculture's contribution towards climate change. The reality is that we already have the knowledge and resources needed, but they are being deployed within a system that is inherently wasteful, exploitative and characterised by inequality. "Six Steps Back to the Land" sets out a renaissance, a future shaped not by fear but by understanding.
***John Turner farms 250 acres in Lincolnshire, and is a co-founder of the Pasture-Fed Livestock Association***

As the world population edges up to the ten billion mark, the question of how people are going to be fed grows ever more urgent. Bigger units! More machinery! the politicians and the corporations say. Colin Tudge is one of the most persuasive, reasonable, hard hitting voices calling for an alternative. This book spells out the steps by which we might be able to get out from under the dominance of the Corporations and the 'big is best' mentality, and restore a farming which will feed 10 billion without poisoning the landscape.
**Tim Gorringe is Emeritus Professor of Theology, Exeter**

In his wonderfully accessible style, Colin Tudge gathers his well-honed arguments about society's miss-directed development of agriculture. We could diminish the major problems of the past while sustaining a happier, healthier and more equitable future for agriculture and thus for society as a whole.
**Professor Martin Wolfe, Wakelyns Agroforestry, Suffolk**

# Six Steps Back *to the* Land

## Why we need small mixed farms and millions more farmers

**Colin Tudge**

Published by Green Books
An imprint of UIT Cambridge Ltd
www.greenbooks.co.uk

PO Box 145, Cambridge CB4 1GQ, England
+44 (0) 1223 302 041



First published in 2016, in England



Front cover illutration by Clifford Harper
Interior design by Jayne Jones
Cover design by Mad-i-Creative

ISBN: 978 0 85784 300 5 (hardback)
ISBN: 978 0 85784 302 9 (ePub)
ISBN: 978 0 85784 301 2 (pdf)
Also available for Kindle.



10 9 8 7 6 5 4 3 2 1

To Bob Orskov, who first opened my eyes to
the realities of agriculture

and to

Ruth, my wife, who turns bright ideas into action

# Other books by Colin Tudge

(The most recent books are shown first.)

*Why Genes are Not Selfish and People Are Nice*

*Good Food for Everyone Forever: A people's takeover of the world's food Supply*

*Consider the Birds: who they are and what they do*

*Feeding People is Easy*

*The Secret Life of Trees*

*So Shall We Reap: the Concept of Enlightened Agriculture*

*In Mendel's Footnotes: Genes and Genetics from the 19th century to the 22nd*

*The Variety of Life: A Survey and a Celebration of All the Creatures That Have Ever Lived*

*Neanderthals, Bandits and Farmers*

*The Day Before Yesterday*

*The Engineer in the Garden: Genes and Genetics from the Idea of Heredity to the Creation of Life*

*Last Animals at the Zoo*

*Global Ecology*

*Food Crops for the Future*

*The Food Connection*

*Future Cook (Future Food in the US)*

*The Famine Business*

**Co-authorships**

*The Second Creation: Dolly and the Age of Biological Control*

*Home Farm*

# Contents

# Acknowledgements

A huge number of people have contributed directly or indirectly to this book over at least four decades - some of whom have disagreed vehemently with most of my ideas, but have helped to clear my own thoughts by doing so. The detractors spelled out the ideas that have brought world farming to its knees, and so helped to throw into sharper relief the ideas that really should be taken seriously.

I am far more grateful though, of course, to the many people who have contributed to the ideas in a positive way. First for me was Prof Bob Orskov, whom I met at the Rowett Research Institute in the early 1970s in my *Farmers Weekly* days, when the Rowett was directed by the great Sir Kenneth Blaxter. Bob, now at the James Hutton Institute, Aberdeen, first showed me how the husbandry of animals must be shaped by their physiology and psychology - how the one must follow from the other. Bob also spends a significant slice of each year promoting peasant farming in what Mahatma Gandhi called 'the Third World' (the PC expression 'developing world' is at best a euphemism and is in many ways misleading) and is showing that the West has at least as much to learn from them as they from us. More recently I have been heavily influenced by Prof Martin Wolfe, plant pathologist turned arable farmer, who is a pioneer of agroforestry and organic husbandry at Wakelyns Farm in Suffolk. I regard them both as gurus (whether they like it or not).

In recent years, too, my discussions with archaeobotanist turned arable farmer-cum-baker-cum-thatcher John Letts have been tremendously instructive (to me). Others who have contributed enormously to my general feel for farming and growing include David Wilson, farm manager at Highgrove in Gloucestershire; Ed Hamer, smallholder on Dartmoor in Devon; David and Anneke Blake, smallholders at Cassington near Oxford; Roland Bonney, co-founder of FAI at Oxford (formerly called the Food Animal Initiative); Tim Waygood, mixed farmer at Church Farm in Hertfordshire; Denise and Christopher Walton at Peelham Farm in Berwickshire; and Catherine and Graham Vint at Hornton Grounds Farm in Oxfordshire. All are showing how mixed farms with very high standards of husbandry including animal welfare can work even in today's political and economic climate (and what might be achieved if conditions were favourable).

I have also been informed and encouraged by many people in various capacities who simply think along the same lines and have contributed to the various initiatives that Ruth (my wife) and I have cooked up over the years (see Epilogue). They include agricultural writer Graham Harvey, who first proposed the Oxford Real Farming Conference, which the three of us launched in 2010; Patrick Krause, chief executive of the Scottish Crofting Federation, who introduced us to the enormously important crafts and traditions of crofting; Jon Rae, Tim Crabtree and Rachel Fleming for enabling me to run a course at Schumacher College, Dartington, on Enlightened Agriculture in 2013 (though it was billed as Agroecology); Michel Pimbert, now executive director of The Centre for Agroecology, Water, and Resilience, for marvellous insights and lots of encouragement; Scott Donaldson of Creative Scotland; Tim Lang, Britain's first professor of food policy at City University, London; Geoff Tansey, deep thinker on the politics of agriculture, who runs the Food Systems Academy with its excellent website; and friends and fellow travellers at the Pasture-Fed Livestock Association, including John Meadley, and farmers John Turner, John Crisp and Will Edwards, who honoured me and Ruth by electing us as lifetime members.

I am aware, too, of my ongoing debt to a variety of people over the years who have helped in many different ways to make our various initiatives work. They include Martin Stanley, who first helped us to set up the Campaign for Real Farming and has been unstintingly helpful since; Peter and Juliet Kindersley, who have supported the Oxford Real Farming Conference (ORFC) for most of its life; Sir Crispin Tickell, who set the tone of the ORFC by delivering its first-ever speech in 2010, and has chaired various sessions since; Patrick Holden, now of the Sustainable Food Trust, and Philip Lymbery of Compassion in World Farming, who have been good friends to the ORFC; to the many people who now help us to run the ORFC, and especially Harry Greenfield; and, from the ORFC's early days, Tom Curtis who set up LandShare, and Sam Henderson who now, with his wife Lucy and friends, runs Whippletree Farm in Devon.

I am also very grateful to our team of advisers to FEA (see Epilogue) and the trustees of the Real Food Trust.

If we want to change the world, which of course is the point, then we must, above all, cooperate: and I am very aware of our debt to many other like-minded organizations worldwide with whom we now communicate. If we all had the time and resources to collaborate more fully I am sure we could summon the critical mass needed to put farming back on course.

Finally, I am very grateful to my editor, Alethea Doran at Green Books, for her very helpful comments and for generally improving my initial text.

Most of all, though, I have to thank my wife, Ruth (aka Ruth West). She combines the roles of companion; guru; creative editor (from proofreading to 'That is rubbish. Start again'); technical adviser and general manager (she is co-founder of all FEA's initiatives, as outlined in the Epilogue, and remains its chief organizer). She also helped to set up the All-Party Parliamentary Group on Agroecology which meets once a month in the Commons or the Lords when parliament is in session.

# Author's preface

In 1974, when I was first beginning to get seriously involved in food and farming, I attended the first United Nations' World Food Conference in Rome - convened by Henry Kissinger after a series of famines in Africa and Asia. One of the root causes, the wisdom of the time had it, was the 'backwardness' of Third World farmers: clearly they needed a sharp dose of Western-style development, including Western-style agriculture with plenty of fertilizers and crops custom-bred to respond to them.

Rising human numbers compounded the problem, for as the English economist-cleric Thomas Robert Malthus predicted at the end of the eighteenth century in his *An Essay on the Principle of Population*, the human population was bound to outstrip the food supply sooner or later. By the mid-1970s there were 4.5 billion of us and it seemed that the Malthusian chickens were finally coming home to roost. Human numbers had been rising by up to two per cent per year - which, because of compound interest, meant that the world population would double in 40 years, increasing four times within the lifetime of a single human being. By the end of the twenty-first century, at that rate, numbers would have increased tenfold to around 40 billion. No one, even the most zealous of technophiles, could suppose that we could ever support so many. Truly the end was in sight. Yet there was worse. For ever since the 1930s the nutritionists had been telling us that above all, human beings need plenty of protein - which must be of the highest class, obtainable only from meat. As a rough rule of thumb, animals must eat about 10 kg of plants to make one kilo of meat. Surely, we couldn't possibly produce enough.

That at least is how things seemed in the 1970s.

Mercifully, as we will see in the opening chapters of this book, the picture now seems very different. Human numbers continue to rise but the rate of increase is decreasing and, says the UN, it should come down to zero by 2050 - meaning that the global population should stabilize at around 10 billion. The nutritionists have revised their calculations and now it seems that although meat is a very considerable nutritional bonus we don't need vast amounts of it, and if push comes to shove, we could get all the protein we need from plants. We should be able to feed 10 billion - in fact could do so fairly easily - and numbers should go down after the mid-century peak, so there is no need to panic after all. Furthermore, we should be able to do what needs doing mostly by building on traditional practices, without the high-tech, high-capital industrialization recommended by the West. Phew.

Yet the crisis is not over. For although it should be perfectly possible to ensure that everyone, everywhere, has good food, the world as a whole is not doing the things that could bring this about. In essence, the people with the most influence - the oligarchy of very large companies known as corporates, and banks, governments like Britain's, and their chosen intellectual and expert advisers - are behaving as if nothing much has changed since the 1970s, and are pursuing the same kind of strategies as they did then. They continue to insist that we need more and more food, and that this can be supplied only by Western-style, high-tech, high-input industrialization - driven these days, as was not quite the case in the 1970s, by the ultra-competitive global market. If high tech and the free market fail to deliver it can only be, the current thinking goes, because the rest of the world has not yet come to terms with them - and that people the world over, especially the poorer people, have failed to curtail their numbers.

So the world still has food problems, just as it had in the 1970s, but they no longer seem intractable - or not, at least, to the analysts who are best informed. It's just that we, the world, following our leaders, are doing the wrong things. The world's present plight cannot be explained in simple Malthusian terms, and there are good reasons to believe that although Western-led science and high tech have a very great deal to offer, over-zealous application can do far more harm than good. But those with the most power continue to apply the same kinds of solutions now as in the mid-twentieth century: high-input, high-tech industrialization on an ever larger scale, all driven by the market and increasingly under corporate control.

Above all at the Rome conference in 1974 I was shocked by the attitude of the delegates. The most powerful nations, including Britain and the US, were not there to ask, 'How can we ensure that everyone has enough to eat?', which I had naively assumed was the point of the conference. They sought instead to defend their own political and economic positions. They seemed anxious above all to demonstrate that whatever was going wrong was not their fault and that the world would be best advised to follow their lead. They, after all, were rich while most of the rest of the world was poor, with India and China still prominent among the desperate. It followed - did it not? - that the rich, the US and Britain and our ilk, must be doing the right things, and the rest must be getting it wrong. Only a few countries - Canada and China were among them in those days - seemed seriously focused on what was supposed to be the question in hand: on how to make sure that everyone has enough food. I was shocked, too, to find that many of the people who were making the decisions that affect all our lives were not well-informed - certainly not about farming, which was and is at the heart of the matter. Only a few farmers were in attendance, as opposed to agribusiness people, and their presence was definitely token. The bona fide farmers mooched about in bewildered convoys.

Taken all in all, *plus ça change*. The people with the most power still present the problems in ways that suggest that the world at large must need their ministrations. However things may appear, the message is that those in charge are doing the right things and ought to stay in charge. 'Consider that you may be mistaken' was Oliver Cromwell's advice to the members of the Scottish parliament in the 1640s, but this is the last thing that the powers that be, then as now, can ever bring themselves to do.

Archbishops and senior scientists alike of late have warned that humanity and our fellow creatures are facing Armageddon - yet if only we did things differently, we and the rest of creation might still be looking forward to a long and glorious future. All the solutions we need are already out there, or so all the evidence suggests. But they are not the solutions suggested and imposed by the world's most powerful people. Those in whom we have placed our trust, the corporates and governments like Britain's and their intellectual entourage, are pulling us in quite the wrong directions. We need to change course radically; not simply by reforms which won't get us where we need to be; nor by wholesale revolution which most people don't want and is far too risky. Rather, we need to bring about a renaissance - a rebirth. We must start again from first principles, though building on the good things that already exist - of which, mercifully, there is a great deal. In particular we need an Agrarian Renaissance. But since the people in power are thinking along quite different lines, we - all of us; people at large - have to do what needs doing for ourselves.

So that's what this book is about. Why we need the Agrarian Renaissance, what it entails, and - most importantly - how we, people at large, ordinary Joes and Janes, can make it happen.

***Colin Tudge, Wolvercote, Oxford, 2015***

PART ONE

# The road to enlightened agriculture

CHAPTER ONE

# What is and what could be

Everyone, everywhere, could have plenty to eat. Farming could and should once again be seen as a desirable, enviable pursuit - assuming its rightful place at the centre of human affairs. As never before, humanity could be at peace with itself and with our fellow creatures and the biosphere at large. We, humanity, might all be looking forward, realistically and with equanimity, to the next million years - when our descendants might draw breath and contemplate the following million. All this is eminently achievable. All that now stands between us and a long and glorious future is seriously bad strategy based on false ideas that happen to be convenient to the people with the most power: ideas rooted in a debased ideology that puts short-term wealth and dominance above all else.

So as things stand the picture is far from promising. The United Nations is telling us that out of seven billion (7,000 million) people now on Earth, nearly one billion are chronically undernourished. Another one billion are chronically overnourished, or at least their diet has drifted too far from the physiological and microbiological comfort zone and makes them vulnerable to a catalogue of diet-related diseases including coronary heart disease, stroke, various cancers and, above all, diabetes. The World Health Organization tells us that the world population of diabetics is now around 350 million - well over twice the population not simply of Wales, or even of France, but of Russia. Obesity, the most obvious first sign of overnourishment, does not seem to be especially dangerous by itself but is associated with many conditions that are, including most of the above.

According to the UN one billion people now live in urban slums, which is almost one in three of the 3.5 billion who live in cities. Yet most of the world's economic strategies, including those of agriculture, seem designed more or less deliberately to drive more people away from the countryside, for urbanization is seen as a sign of modernity and hence of progress. At the same time, biologists conservatively estimate that half our fellow species - say about four million out of an estimated eight million - are in imminent or at least realistic danger of extinction; the most dramatic mass extinction since the end of the dinosaur age, and certainly the most rapid of all time; and, directly or indirectly, agriculture is the main cause.

For good measure, the prevailing economic dogma of the past 30 years - that of the neoliberal, global, 'free' market - has demonstrably made the poor steadily poorer and the rich incomparably richer, while the middle classes have at best stood still. And overshadowing everything, threatening to make a nonsense of all our aspirations and dwarf our present ills, is global warming - now apparently on course to send the world's climate into something resembling chaos perhaps within the next few decades.

All of the above results in large part from the way we farm, and distribute food. Present-day agriculture, or at least the kind that most governments perceive to be modern, and now support with their power and our money, is above all industrialized: high tech (technology based on modern science, as opposed to the kind that emerges from craft, like mediaeval windmills); high capital; and vast in scale. Some statistics, carefully selected, suggest that industrialized farming is a runaway success. Yields in favoured fields and from 'elite' (the most productive) animals exceed the wildest dreams even of 50 years ago and have been achieved with far less labour, which is seen as a prime measure of efficiency. (Official stats rarely deduct the cost of bankruptcy and unemployment, and the suicides that have so often resulted when farmers and their workers lose their jobs as Vandana Shiva discusses in *The Vandana Shiva Reader* [see Resources], or the collateral damage that results from over-simplified husbandry). Yet the global statistics tell us that for all its flashiness and hype, present-day agricultural strategy is failing - if, that it is, we are naive enough to suppose that the point of farming is to provide us all with good food and to take care of the rest of the world.

Those who preside over the status quo are wont to tell us that famines and malnutrition are inevitable, and are likely to get worse - but mainly through the fault of humanity; not because of misguided policy. According to the UN, the human population is on course to reach 9.5/10 billion by 2050, and individual expectations are rising too: in particular, the demand for meat worldwide has doubled over the past 50 years. Hence the idea, now virtually the dogma, which says that the world will need 50 per cent more food by 2050, well within the lifetime of people now in middle age, let alone of our children. Given that the collateral damage even from present-day farming is so enormous, dreams of a long and glorious future seem ridiculous. The appropriate response, one might reasonably feel, is panic.

Yet, as is so often the case, the official line is far from the truth. In some important respects it is the precise opposite of the truth. For while the UN demographers do indeed tell us that human numbers are on course to reach 9.5 and perhaps 10 billion, but they also say that the population should then level out. For the *percentage* increase has been steadily going down over the past 30 years and by 2050 it should reach zero - meaning there should be no further growth. So 9.5/10 billion is as high as numbers should ever get. If we can feed that many for a few decades, or possibly for a few centuries, without terminal damage to the Earth we'll have cracked the food problem for evermore. Furthermore - one of many serendipities - the reasons for the levelling-off are nearly all benign: women *choose* to have fewer children as they achieve liberation and have more options in life, and as people gain greater access to contraception, and as medicine improves and infant mortality goes down, more and more people the world over have less and less reason to fear that their children might die in childhood, and so don't feel the need to have one or two more for insurance.

There is one statistic, too, that turns all the official projections on their head. At the House of Commons on 15 March 2011, Hans Herren, president of the Millennium Institution, Washington DC and co-chair of the seminal 2008 International Assessment of Agricultural Knowledge, Science and Technology for Development (IAASTD) report *Agriculture at a Crossroads* pointed out to the All-Party Parliamentary Group on Agroecology that the world is currently producing an average of 4,600 kilocalories of food per person per day. Given that a high proportion of the world's people are children, the average daily requirement is around 2,300 kcals per day - so we are producing twice as much food energy per head as we really need. In fact, said Professor Herren, 'Right now we produce enough food for 14 billion people'. Fourteen billion is twice the present population and is 50 per cent more than the UN demographers predict will be the maximum. The United Nations Conference on Trade and Development (UNCTAD) made the same point in its Trade and Environment Review for 2013, *Wake Up Before it's Too Late.*

In truth, anyone can confirm Professor Herren's assessment for themselves, using standard statistics that are readily available in standard texts and on the net. Thus according to *The Composition of Foods,* which was originally put together in 1940 by Elsie Widdowson and Robert McCance - and despite many revisions by various editors has been known ever since as 'Widdowson and McCance' - all cereals provide roughly 300 kcals per 100 grammes, and they also contain protein in roughly the proportion that healthy humans generally require. So 1 kg of cereal provides 3,000 kcals, plus adequate protein - which, as you can see from the preceding paragraph, is more than enough to keep a human being going for a day. Hence, one metric tonne of cereal - 1,000 kg - provides enough macronutrient (or at least, enough food energy and protein) to keep three people going for a year. Of course, human beings do not live by cereals alone, but right now we are simply talking about quantity. We can say that each of us requires *the equivalent* of a third of a tonne of cereal per year.

According to the website Statista, which brings together stats from 18,000 sources, the world as a whole produced 2.471 billion tonnes of cereal of all kinds in 2014-15. That is enough macronutrient for 2.47 x 3 = 7.4 billion people - more than the current world population. But the world also produces huge amounts of pulses, including soya, plus tubers such as potatoes and yams, plus oilseeds (soy, ground nuts, palm nuts, coconuts, rapeseed, olives, sunflowers and others) and sugar crops (cane and beet) which are heavy on calories, plus a large amount of meat that is heavy on protein. (Of course, if the meat is produced by feeding the animals with cereal and pulses, then that would detract from the total amount of cereal and pulses available for human beings. But if the meat is produced from grass or swill, with only small amounts of cereal or with none at all, then the meat adds to the human food supply). People also eat a significant amount of wild-caught fish (too much indeed). Add up all of that and, with the aid of Widdowson and McCance, we can see that cereals provide only about half of our total macronutrients.

So the total amount of energy and protein from all the world's food sources is enough to feed 7.4 x 2 = 14+ billion people. We can argue details but we can readily see for ourselves that Hans Herren's and UNCTAD's figures are at least in the right ballpark. The much-bruited idea (it is now virtual dogma) that we need 50 per cent more food by 2050 very definitely is not. But although I have found noone in high places who disagrees with Hans Herren's statistical analysis, and although it is so crucial, I have heard very few people in high places make the point, apart from Prof Herren himself. Meanwhile, the hype - that we need to produce 50 per cent more food by the middle of this century - continues unabated. Scientists in particular emphasize their respect for evidence, and politicians of late have been using the term 'evidence-based policy'. Yet neither allows mere fact to get in the way of commercial expediency. The priority is to maximize wealth and power (just as I perceived was the case at the World Food Conference of 1974). This generalization applies across the board; not just in agriculture. The world cannot improve until people at large begin fully to appreciate that this is the case, and appoint new leaders (insofar as leaders are necessary) with quite different values and a quite different agenda.

In short, the reason why so many people now have too little to eat has almost nothing to do with our ability to grow food. It is caused by waste - much of it deliberately contrived, as discussed in chapter 3; by injustice; and in particular by the lack of food sovereignty (meaning that more and more people are losing control of their own food supply, as discussed below). The continuing emphasis on production - 50 per cent more by 2050 - has very little to do with need and almost everything to do with the perceived need to maximize wealth.

The specific claim that we need to grow more because people at large are demanding more and more meat at least needs serious scrutiny. Food preference, as has long been apparent, is determined at least as much by custom, fashion and general zeitgeist, as it is by innate predilection. Among other things, meat has been marketed with all possible vigour this past half-century, and marketing works, which is why so much is spent on it. 'Demand' is assessed retrospectively - it's a reflection not of what people say they truly want, but of what, if the pressures are right, they can be persuaded to buy, and that is not the same thing at all. More on meat - the real reasons why 'demand' is growing - can be found in Chapter 2.

The present disasters cannot be blamed on the fecklessness and ignorance of humanity at large, as people in positions of influence are wont to suggest. We have not produced more children than the world can support. Neither should we blame the world's farmers. They already produce enough for everyone for all time and they are not mere stick-in-the-muds, afraid of change and slow to respond to the clarion call of high technology. As individuals and as a species we could still be looking forward to a long and glorious future. World peace should not be beyond us. Personal fulfilment for all should be well within our grasp. The Armageddon mentality that now seems all-pervasive should seem absurd; our descendants could still be here in a million years, enjoying a far better relationship with our fellow species than we have now. The present shortcomings result largely, and in some cases entirely, from ill-directed strategy imposed by an elite, which either does not know what it is doing or does not have the world's best interests at heart, and in either case is deeply reprehensible.

The root answer to our ills - not the whole answer, of course, but what lawyers call the *sine qua non* - is enlightened agriculture.

## Enlightened Agriculture

The dream of a well-fed, well-tempered world may seem fanciful but in truth is eminently achievable; and if everyone was well fed, then everything else we might aspire to would become possible - not guaranteed, of course, but possible: world peace, far better health, equitable societies, personal fulfilment, a flourishing biosphere. We need only to design agriculture as if we really *wanted* to provide everyone with good food and take care of our fellow species. We need what I have been calling 'enlightened agriculture' aka 'real farming', defined informally but adequately as:

'Farming that is expressly designed to provide everyone, everywhere, forever, with food of the highest quality, nutritionally and gastronomically, without wrecking the rest of the world.'

There are serendipities all along the way. Notably, enlightened agriculture is founded on just three simple ideas, all of which are widely acknowledged the world over. Of course, no one is suggesting - certainly not me - that by following any particular prescription we are bound to achieve all the outcomes we desire. This is the mistake of ideologues throughout the ages, both religious and secular, right-wing and left: 'do as I say, and everything will turn out OK'. I do suggest, though, that as far as can now be seen, the component ideas of enlightened agriculture offer us the best chance of a long and agreeable future; and in an innately uncertain world, the best chance is the best we can hope for. It's important, too, not to be pre-emptive as modern-day progressives certainly are, as they sweep aside tried and tested traditions to make way for the latest profitable wheeze. The precautionary principles must apply. We must always give ourselves time to change our minds and scope to change direction.

The essential ideas are:

First, the *method* of enlightened agriculture is **agroecology** - treating farms as ecosystems and agriculture as a whole as a serious, positive player in the global biosphere.

The second grand principle is that of **food sovereignty** - which means in essence that all peoples should have control over their own food supply.

Finally, to provide the framework, to ensure that what is done matches the real needs and aspirations of humanity, and is good for the world as a whole, we need **economic democracy** - which uses standard financial mechanisms of the kind traditionally favoured in the Western world to nurture societies that are just and convivial and to take care of the biosphere. Economic democracy achieves radical change by non-radical means: the judicious use of widely-acknowledged, conventional financial mechanisms for the benefit of humanity at large and of the biosphere. The good news is that we can achieve this without economic extremism, whether it be to 'smash capitalism' or to demonize socialism. Although the present global economy is way off course, it contains elements that can be put to good use, or could do with a little adaptation.

Yet there is a problem. The kinds of things that we need to do to ensure a long and pleasing future for humanity and our fellow creatures - somewhere between agreeable and glorious - are not on the agenda of the world's present leadership. Despite the evidence all around us, those in power seem convinced that they are doing a good job, and that no one else could do it better or should be given the chance to try.

In truth, we don't simply need a change of personnel, we need a change of mindset. In practice, if we truly care about the future - of ourselves, our children and our children's children, and other people's children and all our fellow creatures - then we, people at large, ordinary Joes, have to take matters into our own hands. We ourselves have to bring about the changes we want to see. The method is not to ask the powers that be, cap in hand, if they would kindly change their ways, which (at its worst) is the method of **reform**; nor to fight the status quo head on, which would be **revolution**, and would surely fail, with immense collateral damage; but simply to re-create the world we would like to see, *in situ* - which can properly be called **renaissance**.

So that's what the world needs. We need renaissance in general but more specifically an **agrarian renaissance**, applying the principles of agroecology, food sovereignty and economic democracy to establish enlightened agriculture as the global norm; and this renaissance must be led by *us*, people at large.

The seeds of renaissance are already sown the world over, in rich countries and poor. Here and there some of those seeds are flourishing. The big ideas and most of the mechanisms that we need are already in place.

Here and there in villages and suburbs and philanthropic movements, and within commercial companies of a kind that are now considered eccentric, there are thousands of examples of good people doing good things, individually and as communities. We need only to identify who is doing what and draw attention to them; to learn from them and build on what they are doing; and to draw the threads together - to turn the present, scattered dots of worthwhile endeavour into a global network. We need *coherence*, in short. Then we will have lift-off.

We don't need a majority on side to change the world. We just need a critical mass - and I am sure from common observation (and some quantified evidence) that in people's hearts and minds we already have sufficient mass. Now we need to turn the good ideas and good intentions into concerted action. So:

## What does enlightened agriculture entail?

To create enlightened agriculture that is truly robust we must start from first principles and ask the most basic questions that our present rulers have failed to address.

We must ask, first, how much food (and other agricultural produce) does the human species really need?

Secondly, how much food (and other growable stuff) can the Earth's ecosystems reasonably provide us with?

And finally, what is it *right* to do?

Of course there are hundreds of subsidiary questions which we should never stop asking - for instance, what should we mean by '*reasonably* provide'?

Clearly, 'What do we need?' and 'How much food can the Earth provide?' are, in the end, matters of biology, best addressed by science. 'What is it right to do?' is, by definition, a matter of morality. In short, if we truly care about the wellbeing and the future of humanity and of the rest of the world, then agricultural strategy must be rooted in principles of sound biology *and* of common morality.

All this may seem obvious - yet it is not the way our current leaders see things, in governments and big-time commerce, and it is not how farming strategy is devised. Through all the known history of agriculture some people have felt that agriculture is, as the modern expression has it, 'just a business like any other'. Even in Renaissance Italy - even in the Middle Ages and surely too in classical times - merchants were urging farmers, or dominant states were urging subservient states, not to grow food for the people in ways that were sustainable and benign, but instead to raise as much as possible for those with the greatest ability to pay. Grains, oils, booze, spices, tea and many other such desirables were transported in good condition over vast distances many hundreds of years before the time of Christ.

Emphasis on trade as opposed to survival has often led to disaster - including that of the Irish potato famines of the 1840s. People starved even though the barns were full of oats - for the oats were earmarked for English horses. As always, it's a question of balance, for trade carried out judiciously can bring enormous benefits both to the producers and the consumers, and overall is very necessary. Sensible trade, nowadays often manifest as 'fair trade', is very much a part of enlightened agriculture. Inescapably, though, there is a tension, and within historical times there always has been between the need to farm within the bounds of nature and for the benefit of the community at large, and the desire to maximize wealth. The perennial problem is how to resolve that tension; to achieve the sensible balance. In today's world, in agriculture as in all human affairs, the balance is horribly askew. The perceived need to maximize wealth trumps all other considerations.

In 2011, perhaps alarmed by the findings of UNEP's report of 2008 by IAASTD, the British government commissioned its own report, convened by its chief scientific adviser Sir John Beddington. This report, *The Future of Food and Farming*, said that the world needs 50 per cent more food by 2050 and that supplies could be maintained only if each country exports what it grows best and imports whatever else it needs, according to the principle of 'comparative advantage' first suggested at the start of the nineteenth century by the English economist David Ricardo. Indeed, says *The Future of Food and Farming,* countries that refuse to export food should be forced to do so by some higher authority, even in times of famine, just as in nineteenth-century Ireland. For in the most powerful circles, farming is conceived *only* as a business like any other.

Even this view need not in itself be totally disastrous, provided that the business has a moral and social dimension, and is seen both as a means to unleash the spirit of individual creativeness, and to serve society (what's now known as 'social enterprise', of which more later). But in the prevailing, free market economy, known as 'neoliberal' (see Chapter 4) business itself has been reconceived – it is very different now from how it was, say, in 1950s Britain or in the Netherlands of the seventeenth century. Now, business has no in-built morality and attempts to restrain its zeal are seen as heretical or interfering. Restraint for social reasons is deemed to be socialist, and 'socialist' has become a term of abuse – even, since the days of Tony Blair, within the higher echelons of Britain's Labour Party. Those who express concern for the wider world, the biosphere, have been called 'eco-fascists'. What a terrible thing it is to ask rich people not to trash mangroves and the coral reefs that depend on them, when a marina would be so much more profitable! (Or, as I saw recently in Panama, as a place to store containers). Often, though, of course, the 'rich people' aren't just people. They are embedded within corporates that have the legal status of people but have no soul at all. That is a very dangerous condition for the world to be in.

Agricultural strategy in the prevailing neoliberal economy is shaped by the perceived need to make more money, more quickly, than anyone else – for all of us are required to feed our produce and our services into the global market and the market, by stated intention if not in fact, is supposed to be maximally competitive, and money and market share are its only goals. Farmers must compete with farmers from all over the world to make the most money – and indeed must compete for investment with all other industries and profitable pursuits, from nuclear missiles to hairdressing; and in Britain in particular they have to compete with speculators who want to buy their land which at this moment is seen as the safest investment.

To put the matter another way: today's agriculture is shaped not by human needs or the needs of the biosphere, or morality, or even common sense, but by accountancy, in turn perverted by the manipulations of finance capitalism. What is actually *done* in the end reflects what bankers and hedge-funders find most convenient. The whole sorry mess is supported by governments like Britain's, committed to the dogma of neoliberalism, while the Labour 'opposition' has been singing with equal vigour from the same neoliberal hymn sheet.

Neoliberalism says in practice that all goods should be regarded as commodities to be traded in a free global market which in theory (though not in practice) is open to all, from the smallest African farmer to Unilever and Monsanto. As far as possible, its advocates say, the market should be left to do its thing: deregulated. Then what is produced, in what quantities, and how, is determined by the market: how much people are prepared to pay for their goods.

Since the market depends on buyers - consumers - and since the consumers are us, it seems that we control the market and therefore that the market is democratic. But in practice, as is inevitable, the market is dominated by the biggest players, which means the corporates. At the same time the consumers who count most are the ones with most money so in practice the market is dominated by an elite, an oligarchy, of corporates, and it caters primarily for the rich who have the most buying power. The rest, which includes most of the world's population, must squeeze in where they (we) can.

The suggestion that we should farm more benignly, and with less of an eye on short-term profit for the benefit of the dominant few, is written off as unrealistic. The abstractions of finance capitalism are considered more realistic than the realities of the finite world and the claims of human beings and of all other creatures to live agreeably, or indeed to live at all. Apparently the mere creation of wealth, by whatever means and for whatever purpose, will somehow solve all the world's problems. It doesn't, of course, and the resultant mess - the billion malnourished and all the rest - is not surprising at all. The world needs re-thinking on every front (see Chapter 4) but agriculture must be at the heart of all our thinking. Enlightened agriculture is what we need - farming that addresses our real problems. So what, in practice, *is* enlightened agriculture?

## Productive, sustainable, resilient

To feed us all well and to go on doing so we need agriculture that is **productive, sustainable** and **resilient**. This defines the basic task of enlightened agriculture as it should of all agriculture.

We know we can produce enough. We already produce far more than we should ever need, even though so many are still going hungry. We should, though, now and in the future, recognize that 'enough's enough' - in absolute contrast to the dictate of the present, neoliberal economy which seeks to remove all ceilings on production and requires farmers everywhere to produce more and more and more, so as to maximize turnover and hence to maximize wealth. To be sure, land is taken out of production now and again in the Western world, but only to avoid short-term glut - a temporary measure until the market itself can be expanded. The overall thrust is still for more and more. Enough's enough must be the guiding principle - but still, with 9.5/10 billion on board by 2050, enough is a lot.

Present methods, though, are far too destructive. Sustainability must be the priority. Some extremists, of various ideologies, may welcome Armageddon and are keen to bring it on, but most of us would prefer life to continue. Sustainability does not mean that whatever we do now we must go on doing forever. It does mean that we must keep the soil in good heart, as farmers say, so that we can go on producing enough. Indeed, we must keep the whole biosphere in good heart: all landscapes, the oceans, the climate.

Resilience means that however we farm, we must be able to change direction as the conditions change - as change they do and surely will. Notably, the climate seems likely to change very rapidly indeed, and in any one place it is likely to fluctuate as we have seen in Britain on a relatively minor scale over the past few years.

So how can our agriculture become productive, sustainable and resilient? We might sensibly begin by looking for precedents - models - and these are provided by nature itself. For nature has been commendably productive for the past 3.8 billion years, which is certainly sustainable enough; and it has managed this even though conditions have veered spectacularly - from ice, pole to pole (or very nearly), to pole-to-pole tropics (almost). Indeed, the climate has flipped between extremes many times in the history of the world while the bits of land that form the present continents have migrated all around the globe, interrupting and creating new ocean currents, and adding to the climate's complexities. So nature has certainly proved resilient, too, in putting up with all this. Of course, nature does many things that are not desirable - over time it tends to erode entire continents and sudden wipe-outs on the small scale and the large are almost frequent - so we should not emulate it slavishly. But it is certainly reasonable to ask how nature achieves so spectacularly the things that we would like to do ourselves.

Nature seems to have five main tricks. First, it is wonderfully **diverse**, with millions of species overall and many different species and sub-types - typically thousands or tens of thousands once we include those that live in the soil - in any one ecosystem; and each species is likely to be very various, genetically and physically. This isn't invariably so, for some wild populations are surprisingly inbred and many plants, in particular, are clones - genetically identical. Some wild ecosystems too *seem* to be remarkably un-various - like the vast boreal forests of Canada with only 10 dominant species of trees. But then, those forests had to begin from scratch after the last ice age had wiped the slate clean just a few thousand years ago, so they haven't had time to diversify; and the soil microbes and fungi in the boreal forest, which are far more versatile and mobile than the trees themselves, are already hugely varied. Variety is the norm, and it has tremendous implications. It makes each kind of creature far less vulnerable to parasites - a general term that includes both pests and pathogens.

No one parasite finds enough hosts of the right species in any one place to get a proper foothold - and since all the individuals in any one wild species are ever so slightly different, it won't always find it easy to hop from one host to another. So it is that fungal diseases like wheat rust produce prodigious quantities of spores, and in a field of domestic wheat they can run riot. But in the wild, where the wild grasses are highly various, wheat rust needs all the fecundity it can muster just to ensure the next generation.

Secondly, the different members of any one ecosystem interact - to some extent competitively, but on the whole **synergistically**. Between them the many different species exploit whatever provender their environment has to offer far more efficiently than any one species could do alone. This has been shown in the laboratory, with assemblages of microbes in nutrient solutions; and on the grander scale where, for example, the many different kinds of African antelope, plus zebras, buffalo and elephants, exploit the grasses and trees of the savannah far more thoroughly than any one alone would do (and the plants have adapted accordingly).

These interactions lead to **cyclicity** - the third great trick. Everything is recycled. One creature's excreta is another's sustenance - for example, the dung of elephants (and other herbivores) provides food for dung-beetles - and dead animals feed plants - dead dung-beetles feed fungi and microbes, and so on. Nature can be leaky but all in all its miserliness is wonderful - and even the stuff that leaks from natural ecosystems finds its way back into the cycle eventually. Some examples are given in Chapter 5.

Fourthly, wild ecosystems on the whole are **low-input**. Sometimes some favoured spot is richly nourished - in estuaries, and wherever animals gather in large numbers - and then various specialist plants and algae cash in on the benison, and animals cash in on the plants and algae, like all those vast flocks of estuarine waders. But most of the natural environment most of the time is of low fertility, at least compared to modern farms - and most wild plants prefer it that way. Certainly, nature never uses fossil fuel to enhance its energy supply, or produces artificial fertilizer. It is powered entirely by the sun (which also drives the winds and rain) and the gravity of the moon (which drives the tides), and to some extent by geothermal heat; and fertility comes from natural nitrogen fixation and recycling.

Finally, **nature does not cultivate**, at least on the macro-scale. The ground is sometimes stirred up, for example when trees are blown over, or herds of big animals congregate, but on the whole the soil is left undisturbed. There is micro-cultivation, however, by deep penetrating roots that can split rock, and by a host of small animals among whom earthworms are outstanding, as Charles Darwin pointed out 150 years ago.

In short, nature succeeds as well as it does because it is enormously diverse, is integrated, is low input but with a marvellous ability to recycle, and prefers on the whole to leave soils undisturbed. So how can farmers emulate such qualities?

Well, **diversity** means many different species on any one farm, both of animals and crops, and genetic diversity within each. The fancy word is **polyculture** - or, more simply, the mixed farm. Ancient varieties of crops were and are 'landraces' - populations selected in situ over many generations of plants that happen to do well in their local conditions; and landraces generally are highly various. Livestock too: I remember being shown a traditional group of breeding rams, all from the same breed but all ever so slightly but noticeably different. The breeder made a virtue of this: 'You don't want them all the same,' he said.

In a sensible world, you don't. Cloning in the short term could be profitable (although various biotech ventures so far have lost a lot of people a lot of money) but from a biological point of view it's a seriously bad idea. Ultra-modern breeders, though, are keen to replicate the most productive - elite - individuals, not least by variations on the theme of cloning, as commended in passing in *The Future of Food and Farming* report. The resulting risk of disease is countered with antibiotics and germ-free environments and other high-tech wheezes, all of which are potentially lucrative and contribute to GDP, so what the hell?

**Integration** means everything on the enlightened farm, or indeed in a well-run traditional farm, feeds in to everything else. This indeed is the essence of agroecology, which is the running theme of Part II. Mixed, integrated farming perhaps reached its height in Britain in the 1950s, with pigs playing a big part - feeding on crop residues or on turnips grown as break crops, and whey from cheese-making. The traditional mixed farms of Southeast Asia are among the pinnacles of human achievement - carp and ducks in the paddy fields among the rice; horticulture on all the higher ground, with cabbage, yams and a great deal else besides; water-buffalo doing much of the work and adding their dung to the mix; a pig in every yard; chickens everywhere; human excrement pressed into service too (that has got to start happening on a bigger scale, not least to conserve phosphorus); and miscellaneous trees serving miscellaneous purposes.

Such systems produce a huge diversity of nutrients and if there is anything resembling justice or fair shares then nutritional deficiency among humanity at large should be inconceivable. But when traditional diversity is replaced by monocultural rice, we see, for example, vitamin A deficiency which blinds tens of thousands of children every year, commonly followed by death. The modern antidote to this is not to revert to protective diversity but to supply Golden Rice, genetically modified strains of rice fitted with genes that produce beta-carotene, the form in which vitamin A is ingested. Thus high tech is used not for the benefit of humanity but to patch up the flaws in modern strategy.

Of course, complex, skills-intensive farming with limited machinery is often hugely demanding and, if the politics is harsh, it can lead to great hardship. A Chinese saying goes 'Peasants have their faces to the earth and their backs to the sun'. But every kind of working environment that was ever devised *can* be made nasty. At their best (so my own observation suggests), traditional mixed farming systems with people working together but with time to themselves can be socially unimprovable: true communities. They are the natural state of post-ice-age humanity.

Low-input in practice means **organic**. At least, inputs on organic farms can sometimes be very high - immediately after the field is manured, for example. But organic farmers make do with what nature itself provides. In particular they do not make use of artificial fertilizer. It isn't obvious that all farmers need to follow all the minute rules of, say, Britain's Soil Association, and get themselves officially certified. But organic should certainly be the default position of all farming - what all farmers do as a matter of course unless there is an overwhelming biological or social reason to do something else; and the Soil Association and others deserve our undying gratitude for their work in sustaining and improving organic methods.

Finally, **minimum** (or **zero) cultivation** implies as little digging and ploughing as possible. This can be hard to achieve without the help of industrial chemistry, especially in the constant battle against weeds, but it is certainly to be aimed for.

The general term for farming that seeks to emulate nature is agroecology; and that is the *method* of enlightened agriculture. This is the subject of Part II.

So the enlightened farms we really need - productive, sustainable, resilient - should be polycultural (with lots of genetic variation), integrated, organic, or very nearly so, and with cultivations kept to a minimum.

The political, economic and social consequences are huge, for enlightened farms by their nature must be highly **complex**. Complexity requires plenty of skilled input - so the farms we need must be **skills-intensive**. Skills-intensive doesn't just mean labour-intensive. Enlightened farms need plenty of farmers who should take advantage of all the technologies that can make their lives easier, including traditional tools but also, sometimes, the highest of high technologies; we're not talking slaves and bonded labour to do the work of tractors. If enterprises of any kind are complex and skills-intensive then, in general, there is little or no advantage in scale-up. So enlightened farms in general should be **small-to-medium sized**. 'SMEs' is the technical term: small and medium-sized enterprises. As we will see in Part III, the SME is a vital concept.

In summary: enlightened farms, the kind we need, should in general be maximally diverse (mixed, polycultural); low input - organic farming is the default position; tightly integrated, so that everything feeds into everything else; therefore complex; therefore skills-intensive - with plenty of farmers; and therefore, in general, small-to-medium-sized. All this is the precise opposite of what is now recommended in official circles: high-input, minimum to zero labour monoculture on the largest possible scale. This is known as industrial agriculture, although, since it is now driven by the dogma of neoliberalism, it could be called 'neoliberal-industrial' (NI) agriculture.

Defenders of the status quo are wont to argue that enlightened agriculture is simply idealistic; that we need above all to be hard-headed, and that this must lead us down the high-tech industrial route. But although enlightened agriculture is rooted in principles of morality and indeed of metaphysics, and although it has great respect for traditional practice (as will become apparent), it is shaped at least equally by modern biology. Overall, enlightened agriculture is pragmatic; a straightforward attempt to do what is necessary and possible. By contrast, the neoliberal-industrial kind is driven by the largely unexamined beliefs that money per se is the prime desideratum; that high tech means progress; and that progress, thus defined, is ipso facto good. In short, the status quo is pure ideology, though it is neither sensible nor pleasant.

I did not invent enlightened agriculture. I invented the term, but its ideas come from other people: farmers; scientists; economists and political thinkers of the kind who recognize that economics and politics must be rooted in morality. History, good science, common sense and common humanity suggest that the general approach of enlightened agriculture is *right*. Here we encounter a giant serendipity. For we have often been told that if the human race is to survive in large numbers for more than a few more years, then we must all learn to eat austerely; indeed, that we must all become vegans.

Again, this is profoundly untrue. As we will see in the next chapter, the produce from enlightened agriculture is precisely what is needed for all the world's greatest cuisines. We don't need to be ascetics. We just need to relearn how to cook.

CHAPTER TWO

# The future belongs to the gourmet

Many of the big ideas that now frame world strategies of food and farming, dinned down to us from on high, are just plain wrong; rooted either in ignorance (if the perpetrators of those ideas don't know that they are wrong) or lies (if they do know, but are content to spread untruths).

Among those untruths, and in some ways among the most pernicious, is the notion that we, humanity, cannot be fed sustainably unless we all learn to eat austerely. Yet the kind of farming that can produce all the food we need - rooted in the principles of agroecology - provides us with precisely the foods we need, in the appropriate proportions, to meet all the criteria of modern nutritional theory precisely; and the foods that nourish us best are also the basis of the world's finest cuisines. In other words, if we want to provide ourselves with the kind of farming that really can keep us in good heart, we just need to take food seriously - relearn how to cook and restore food culture to the centre of everyday life. In short, the future really does belong to the gourmet. Here is yet another giant serendipity. Here again we see that all that stands in the way of a glorious future is the obduracy and stupidity of the people we have allowed to take charge.

So how do I justify these comments?

## Plenty of plants, not much meat, and maximum variety

If we farm according to the principles of enlightened agriculture, rooted in those of agroecology, then we will use all the best, most suitable farmland for arable and horticulture - because arable in particular produces the most food energy and protein per unit area; and horticulture fills in many of the nutritional and gastronomic cracks, and with tender loving care can be amazingly productive. But agroecological principles tell us that all farms and smallholdings should be as mixed as is practical and so almost all of them should include some livestock. Thus the main thrust of our agriculture would be to produce a great many plants - in general, the staples on the arable scale, and fruit and vegetables on the garden scale (see Chapter 9: Horticulture). Enlightened farming would also produce some animals, helping grass to grow by grazing it back and encouraging all plants with their manure.

We need not be content with the livestock that are intended mainly to balance out the crops on mixed farms. In the world at large only about a third - at most! - of the designated farmland is really good for arable. The remaining two-thirds of the world's recognized farmland is grassland and woodland where arable may be possible with greater or lesser degrees of heroism but is generally not at all easy. But animals are more than happy to graze the grass and browse the woods. It is usually judicious to feed *some* arable crops to livestock to tide them over the bad times and because, in practice, in order to be sure of producing enough staple food in any one year, it is prudent to aim at least for a modest surplus. So we would, if we practised enlightened agriculture according to the principles of agroecology, produce at least *some* meat - quite a lot of it in wet and grassy places like Britain - though globally we would produce a lot less than we do now, when more than a third of the world's staples are fed to livestock. Again, in accord with the principles of agroecology and out of respect for the biosphere, we would as far as possible leave the natural grassland and woodland 'unimproved'. Our livestock would feed largely or mostly on the vegetation that nature provides. Ecologically, cattle and sheep (and a few others) would simply substitute for the wild ungulates that would have been there in the deep past. In pre-ice age Britain, pre-ice age grazers and browsers included wild cattle (the aurochs), plenty of deer, horses, and elephants in the form of mammoths. The wild pigs stuck mainly to the open woods.

In modern enlightened agriculture both the arable and the horticulture should be as varied as possible: genetic uniformity must as far as possible be avoided. The natural pastures on which the cattle and sheep feed would naturally be highly various, often with scores of species in every hectare. Many of those wild plants would be culinary herbs - the original species from which the cultivars are derived. Some of the aromatics that they contain - a few molecules out of the whole - would find their way into the flesh of the animals that graze on them. It is also permitted to take a small and sustainable cull of wild creatures, including rabbits, deer and pigeons, and at least some fish. All edible parts of all animals should be eaten. Cow heel and rabbit livers are delicacies.

Thus we find that enlightened agriculture, following the ideas of agroecology, would produce plenty of plants from arable farming and horticulture; not much meat - or less, at least, than is now the custom in Britain or the US; and a great deal of variety. In fact the whole output can be summarized as 'Lots of plants, not much meat, and maximum variety'.

Now we encounter two huge serendipities. First, these nine words, 'Lots of plants, not much meat, and maximum variety', encapsulate all the most convincing nutritional theory of the past 40 years. Secondly, this statement describes the core structure of all the world's greatest cuisines. In other words, farming that is designed according to simple biological principles to provide human beings with enough to eat, sustainably, also supports the best possible nutrition and the world's finest gastronomy. This simple truth should be written in six-feet-high letters over the desk of everyone who has anything to do with food policy or with education. Instead, as is so often the case, this obvious truth is precisely at odds with what the makers of agricultural strategy, and quite a few educators, seem to believe. So:

## The best of nutritional theory

I have been interested in food for a very long time and witnessed the shift in nutritional advice from fairly close hand over the past 50 years. In the immediate post-war years, in Britain, doctors and educators, including experts on children's television, emphasized the need for high-quality protein - which was taken to mean the flesh of animals. Only meat, it seemed, contained enough protein, with the right proportions of all the necessary amino acids. In the 1950s few seemed to feel any great threat from saturated fat so the fact that a high-meat diet meant a high-fat diet was not considered especially important. Carbohydrate in general tended to be written off as empty calories and what is now called dietary fibre was downgraded to roughage: good to arouse recalcitrant bowels and heavily promoted by school matrons but not taken seriously in hard-nosed medical circles. There was, however, significant emphasis on vitamins.

### *Protein and fat*

The emphasis shifted radically in the 1970s. Indian nutritionists in particular pointed out that human beings do not need anything like as much protein as was being recommended (for if it were not so then most of the human race would already have starved) and that plant proteins, including those of the cereals, pulses and potatoes, were for the most part perfectly adequate on their own. That is, if a person has enough wheat or rice to provide all the energy they need, then they will also have enough protein. Nutritionists also pointed out that whereas cereals such as wheat do contain less than the ideal proportion of the essential amino acid lysine, pulses contain more lysine than is needed - so the two together make a perfect balance. The theme of cereal plus pulse runs through all cuisines - chapatti or rice with dhal, tortilla (maize) with frijoles (kidney beans), beans on toast, and so on. In some Indian breads, too, chickpea flour (besan) is mixed with rice or wheat flour - and the result is very pleasing. The main nutritional role of animal-based foods, so it seemed, was to supply micronutrients of a kind that are hard to get from plants in adequate amounts, such as vitamin B12, calcium and zinc. But we could get what we need of these from only a small amount of meat and offals.

At the same time, attitudes to fats changed and became much more refined. First, in a broad-brushed way, nutritionists began to emphasize the possible or probable role of fats in promoting atherosclerosis (in which arteries become clogged with fatty plaques), leading to coronary heart disease. Later they perceived that high-fat diets promote some forms of cancer, including breast cancer (not least because many carcinogens are fat-soluble). They also started to draw a clear distinction between saturated fats, found principally in animal fat - and particularly in animals that had been raised quickly on a diet high in concentrates, meaning cereals and pulses; and unsaturated fats, found in leaves, seeds and fish oil, and also in the muscles of cattle and other livestock that had grown more slowly on a natural diet of grass and browse (the leaves and twigs of trees and shrubs). Saturated fat seemed to promote heart disease, while reasonable amounts of unsaturated fats - olive oil, sunflower oil, fish oil - seemed if anything to be protective, and essential for normal growth and especially for the development of the nervous system, including the brain. In general, there was a radical and rapid shift in emphasis and advice away from a high-protein, high-meat diet that was also rich in saturated fat, towards a more plant-based diet (but with fish considered especially desirable) with a far higher ratio of unsaturated fats.

In truth, the story is proving to be far more complicated than it seemed in the 1970s. Many now doubt that the fat of terrestrial animals (as opposed to that of fish) is intrinsically bad, or that it is bad simply because it can be highly saturated. Unsaturated fats have now emerged as a very mixed bag, and some are far more beneficial than others; and how they affect us depends in large measure on how they have been cooked or otherwise processed. It's also clear that the composition of fat from cattle and sheep depends very much on what they have been fed; in general, the fat of pasture-fed animals contains a far higher ratio of unsaturated fat than those fed on a richer diet with plenty of grain, and the spectrum of different unsaturated fats in a pasture-fed animal may be positively beneficial. In Britain at least, you can identify beef and lamb that has been raised exclusively on grass and browse by looking for the 'Pasture for Life' label, issued by the Pasture-Fed Livestock Association (PFLA).

It seems, too, that saturated fat *per se* may be far less damaging than was once supposed. What matters perhaps at least as much or even more is the way the meat, with its fat content, has been processed; and various components of the flesh itself, the red meat, also seem to exert a considerable influence. For the time being we have warring opinions. There is a lot of work to be done before we have clarity.

For my part, being of conservative mien, I am sticking for the time being to the late-twentieth-century idea that the relatively low fat diets of a traditional Mediterranean or Asian kind are healthier than the high-fat diets of the North, and that polyunsaturates like sunflower oil and fish oils, or mono-unsaturated olive oil, are to be preferred; and that the fat from animals fed on natural diets (which in cattle means grass and browse) is likely to be much better for us than fat from animals raised on concentrate (rich in cereals and soya). I don't worry too much about the fine details which in large part have still to be worked out.

But one broad generalization is growing stronger and stronger: that food should be as *natural* as possible. I do not suggest, as some do, that food should ideally be left uncooked but we surely should avoid gratuitous interventions, including all those additives that are introduced simply to add colour or lengthen shelf-life or provide the illusion of bulk or enhance mouth feel (a ghastly expression) and all the rest. *Haute cuisine*, as great chefs like Raymond Blanc emphasize, is rooted in peasant cooking; and the essence of peasant cooking is its simplicity - fine technique and tender loving care lavished on the produce of the fields and the woods and streams. (Raymond Blanc is French and the Italians are wont to say that the French over-elaborate. But I will leave them to fight that out.)

### *Fibre*

In the watershed 1970s, too, there was a radical change in attitude to dietary fibre. Fibre is compounded of plant cell walls, made primarily of cellulose but also from a host of other accompanying compounds including hemicelluloses and pectin. Fibre doesn't, it turns out, simply provide bulk and scour the gut like a pumice stone. Instead it partially breaks down in the colon and interacts with the host of micro-organisms that form the gut microflora, and this in turn has profound effects on the physiology as a whole. This was largely what prompted the shift of name - from the dismissive 'roughage' to the grander 'fibre'. Low-fibre diets, for example, were shown to promote gallstones (because they affect the gut microbes which in turn influence the re-absorption of bile salts).

### *The emerging science of cryptonutrients*

Finally, in the 1990s we began to hear about 'nutraceuticals' or 'functional foods'. Whatever the name, this is a heterogeneous class of agents whose status is somewhere between that of tonics and bona-fide nutrients. They are more like tonics insofar as they contribute to health, or are presumed to - and yet are not absolutely vital to survival as the recognized vitamins are. As with bona-fide nutrients, the benefits they bring tend to be long-term; they do not generally serve, as tonics are commonly supposed to do, as short-term pick-me-ups. Already commercialized are plant sterols, which seem to lower blood cholesterol, and, for example, are included in at least one widely available margarine. On the market, too, are 'friendly bacteria' which produce various agents that are meant to sort out the gut (although the output of microbes is diverse in the extreme and what exactly does what is very hard to pin down). In short, the idea that these somewhat shadowy agents exist and may be important has been picked up both by the pharmaceutical industry, which prefers the term nutraceuticals, and the food industry, which favours the alliterative 'functional food'. I prefer the term 'cryptonutrients' , which seems to cover all eventualities.

Cryptonutrients are still largely hypothetical and exactly what might be included in this category is mostly unknown, so some would write them off as mumbo jumbo: 'not enough evidence!', they say. Indeed, the idea would probably have been dropped already if various commercial companies had not seen their financial possibilities. But there seem to me to be good biological reasons to take them seriously. After all, all living organisms are master organic chemists. All produce a huge number of strange molecules that serve a vast range of purposes - not least as toxins and repellents to ward off parasites. In particular, microbes, fungi and plants of all kinds produce an entire pharmacopoeia of recondite molecules - and animals, particularly omnivorous animals like us which in principle can and do eat almost anything, would over evolutionary time be exposed to many thousands of them. Our ancestors would have had to adapt to whatever was around, and adaptation often meant acquiring the means to detoxify. Evolution, above all, is opportunistic. It is easy to see how some of the mechanisms that evolved for the purpose of detoxification could later be adapted for more positive ends (just as some ancient bacteria first learnt to cope with oxygen, which is very dangerous stuff, and from that developed the skills of aerobic respiration). When this happens, the erstwhile toxin emerges as a benefit. Then it would qualify as a cryptonutrient.

We would not expect to be able to identify all the potential cryptonutrients. There could be many thousands. Neither, therefore, could we ever hope to identify all their effects - because many of those effects would be low key and drawn out (lowering blood cholesterol perhaps reduces the chances of heart attack but only over a long time); and because they would often manifest as 'cocktail effects', with several different cryptonutrients working together. In this way, a few thousand different agents potentially have many millions of effects - far too many to trace and analyse in the lifetime of a human being, or even in the total lifespan of humanity. So the presence and effects of cryptonutrients can never be pinned down exhaustively. Yet there are good theoretical reasons, and some data, to show that they do exist and that their effects are real - with the expectation that over time they may prolong life and raise our general performance, including our resistance to disease. In case you find this thesis implausible, look at it the other way round. It surely would be close to a miracle if the shortlist of known vitamins that nutritionists have identified these past few centuries represented vitamins in their entirety, given what life in its infinite variety has thrown at us and our ancestors and our bodies. The vast, growing, but still largely hypothetical class of cryptonutrients at least represents a first attempt to tie up some of the loose ends.

All this contributes to the idea that our food should be varied: the more varied the diet, and the more it includes unusual plants, and fungi, and bits of animals in addition to muscle, which is what we call meat, and the more various the diet of the animals that supply the meat, the more likely we are to imbibe cryptonutrients - with a real possibility of long-term benefit.

Whatever way you look at it, in short, 'plenty of plants, not much meat, and maximum variety' is sound nutritional advice. But the serendipity does not stop there.

## The world's greatest cuisines

If the nine-word catchphrase (plenty of plants, not much meat, and maximum variety) added up to austerity then we might face a dull future, compared to one with burgers and fried chicken with their warm savoury niff of much-cooked fat, spiked with onions and gaudy with ketchup, that now abound in every high street. But it does not. *All* the world's greatest cuisines, all of them rooted firmly in traditional cooking, are based on cereals (wheat, rice, maize). All use a huge variety of vegetables and whatever fruits are around. Some make great use of fungi. All use meat only sparingly - as garnish, as stock and just occasionally en masse in special feasts - and all make use of *all* of the animal from the snout to the hoofs. All are immensely varied - everything that's around is incorporated into the world's best cooking, including a wide array of spices and herbs; and the food plants are bred for flavour (implying biochemical richness) rather than as now for yield and shelf-life, while the animals are raised at a modest pace on a varied diet and so are far more diverse, biochemically, than their industrial stuffed-to-the-gills descendants, raced pell-mell from conception to abattoir. All make extensive use of fermentations in many different ways, including the many arts of pickling; and the microbes and fungi that provide us with pickles and booze and hundreds of variations of cheese and yoghurt are the greatest biochemists of all.

The more southerly Chinese cooking, including the finest banquets, is based largely on rice - garnished with whatever is around. Indian cuisine is generally spicier than Chinese and some of its finest recipes eschew meat altogether. Nothing surpasses Turkish or Lebanese cooking - with rice and various breads, olive oil, coriander, mint, almonds, pistachios, apricots, and some of the world's finest green vegetables, and with animal fare only when available - fish caught inshore from small boats, the odd goat, yoghurt, and so on. In northern Europe, including Britain, and in part of the US, the diet was often more meat-oriented than in the East and South because the growing season for vegetables is shorter and because Britain, at least, has plenty of rain and hills - good sheep and cattle country. But British cooking too - I would say at its finest - is heavily plant-oriented, with plenty of bread and pastries. Consider the traditional meat-and-two veg, and the great dishes of the north, such as the unsurpassable potato pie of Lancashire. Yorkshire pudding is a great invention too; and the haggis, based on oats; and, down south, the traditional Cornish pasty - pastry, potato, and turnip, no more than flavoured with meat.

In summary, we have nothing to fear from a future in which we again took cereals and horticulture seriously and raised modest numbers of animals on diets and in conditions far better suited to their biology and psychology. Proper cooks, to be found the world over not only in the grandest restaurants but in millions of anonymous kitchens, would be in heaven. Truly, the future belongs to the gourmet. At least, if it doesn't, then we won't have a future at all. If all that's needed is traditional cuisine, why have we allowed it to be compromised? Why, in Beijing, is the host of traditional restaurants giving way to chains of fried chicken and burgers? Why have (almost) all the pie and mash shops gone from London?

Why, in particular, do we now squander more than a third of the world's cereal, our staple food, and wreck the world in the attempt, to produce meat that we don't need and is far too rich in saturated fat, when we could produce all we need and of a higher quality just by raising livestock on grass and browse, in places where it's hard to grow cereals? Why are we making life so hard for ourselves, and causing such destruction and cruelty, when there is so obviously a better way?

## Why the current emphasis on meat?

Beyond doubt, the reasons the world has placed such emphasis on meat in the last few decades in particular are many and complex but the main ones fall under four headings.

The first is a genuine, well-meaning mistake. In the 1930s nutritionists set out to find the causes of malnutrition in Africa and particularly of kwashiorkor, the peculiar condition in which children swell up - not with fat but with oedema (a build-up of fluid) - while their skin becomes dry and their hair turns reddish. The specific cause, the research concluded, was lack of protein. That's when the idea started that people in the world at large were likely to be short of protein in general and needed animal protein in particular. World War II interrupted the research and the idea lived on through the 1950s to the 1970s, when it finally became clear that we don't need vast amounts of protein at all, except perhaps occasionally under severe physical stress (for example after major surgery). Children with kwashiorkor, it turned out, are not specifically short of protein. They are short of food in general. They are so short of food energy that their bodies burn their own muscles as fuel and the resulting biochemical turmoil leads to pooling of fluid and hence to oedema. By the time this was realized it was already widely believed that meat (and milk and eggs) were all that were really worth eating. So the post-war emphasis on livestock was in large part based on a scientific mistake. But the mistake was understandable and the nutritional advice that followed was, to a significant extent, a genuine attempt to put things right.

The second reason has more to do with perceived wants than with needs. When people are very poor they typically don't eat much meat because they can't afford it. As they grow richer they tend to eat more. America in the 1930s experienced the mother of all depressions and many had too little food of any kind and certainly very little meat. Today the Chinese people are emerging from decades if not centuries of terrible deprivation, including mass famines even within living memory. What food there was would surely have been a million miles from what Chinese cuisine could be, and meat of a tolerable kind must have been rare indeed. In such cases meat becomes more than a food. It seems to provide proof that the horror is over, or at least the worst of it. It becomes a status symbol. Only a few human beings actively seek to dominate everyone else but everyone needs status.

The third and probably the biggest driver of meat-oriented agriculture and meat-oriented cuisine is commerce, which while vital (see Chapter 13) can all-too easily degenerate into scam. As outlined above, agriculture these days is geared not to the needs or true desires of humanity or the wellbeing of the biosphere, but to the maximization of wealth, at least for those who have most influence. But to maximize wealth it is necessary to maximize turnover - and the market for food is all too finite. There is no limit to what people can spend on houses or clothes or works of art but there is certainly a limit to what they can eat. Politicians, industrialists and their chosen intellectual advisors continue to tell us we must focus on production, but in truth it is all too easy to produce surpluses, as demonstrated by the cereal mountains and the milk, wine and olive oil lakes of Europe in the 1980s. Agriculture has become 'agribusiness' - farming primarily for profit - and, above all, market ceilings must be raised or preferably removed. Arable farming is the core of agribusiness. It is the prime source of our staple foods and the principal consumer of fertilizers, herbicides and pesticides, and now of the biotech industry, which are the big money-spinners, particularly in countries like Britain which exports high tech. So the problem for agribusiness becomes - how do we create an open-ended market for arable crops, which mainly means cereals but also, these days, means pulses in the form of soya?

Certainly not by promoting traditional cuisine. Well-off Turks, Chinese, French or Italian people already consume all the rice, bread, and pasta they can manage. More lucrative by far is to feed as much cereal as possible to livestock and then sell the meat. When even the market for meat becomes saturated - well, just throw away most of the animal, or put it in pet-food or fertilizers, and sell only the steaks and chops. These days there is an even easier route: burn the surpluses of grain and call it biofuel and pretend that this is good for the environment. Either way, the emphasis on production has nothing to do with need and everything to do with money, at least for the dominant few. All of agriculture and hence the life of humanity and of the whole biosphere are thrown off course. But a few grow rich, and those few call the shots.

Finally - which may seem trivial but isn't - meat has become fast food. Good traditional cooking requires some skill and dedication but anyone can put a steak under the grill. 'Slam in the lamb', said a recent ad campaign in Britain. People do tend to go for easy options and it is certainly in the interests of those who seek to dominate that we should do so. Self-reliance and traditional food culture are the enemies of oligarchy. Hence, behind the apparent zeal for meat in newly rich societies lies enormous and effective commercial pressure.

So when the Americans and Chinese and everybody else began to emerge from poverty they upped their intake of meat. Some of this surely has to do with genuine desire, because meat, most people agree, tastes nice. But much of this carnivorous zeal has to do with relief - at last the bad times are over! - and even more with fashion; and also, crucially, with tremendous commercial pressure.

We should ask - it's an important question - how much meat would people opt for in the absence of all the confounding variables? How much meat would most of us eat if all the options were laid out before us and we were truly free to choose? There is plenty of reason to think that most of us wouldn't eat much. The old Chinese mandarins who could afford anything they wanted used to leave the flesh of the duck to the servants, and eat only the crispy skin. They wrapped it in tiny wheat roll-ups with various garnishings and it became Peking duck. In modern affluent Germany, or Boston or California, or indeed north London, vegetarianism is chic. At the other end of the spectrum, chimpanzees like to feast on meat - but not every day: and when they have a monkey to feed on they alternate mouthfuls of meat with mouthfuls of leaves. In truth, human beings are natural omnivores and in general, left to ourselves, we do like meat but not a great deal. On the whole, enough to add flavour and texture will do.

The great cuisines reflect this in-built predilection. They were devised not by the food industry or even by celebrity chefs but by people of the kind we call ordinary, over many centuries. They demonstrate the genius of humanity at large.

As always, if we seriously care about the future, we need a radical shift of emphasis and of modus operandi; and as always, we have to shake off the pressures from on high and take matters into our own hands. The collective wisdom of humanity far outweighs the instant solutions and commercial expedients of the powers that be.

CHAPTER THREE

# Why don't we do things that need doing?

Despite its obvious advantages, and despite the evidence, enlightened agriculture is not the global norm. In fact there are very few truly convincing examples of it. Traditional farms often come close to it in structure but for the most part they are under-endowed and cannot perform as well as they could and should; while the industrial kind that are thought to represent progress and are backed by the government-corporate coalition, with our money, are almost the precise opposite of what humanity and the world really need: profligate, unsustainable, and using the wrong techniques for the wrong ends. The trouble is that farming that is guided by biology, and morality is less profitable in the short term than the industrial kind - at least within the present economy. Even more to the point, enlightened agriculture does not concentrate wealth and power but spreads it around among many small farmers and the community at large. This is no good for those who aspire to take their place among the ruling oligarchy or who believe that the world can only be run by an oligarchy which ought therefore to be kept in place.

But the real reason why things are as they are is not made explicit. Instead, defenders of the industrial status quo bring five other arguments to bear. Let's look at these arguments one by one.

First, say the apologists, small mixed farms cannot hope to provide enough food for seven billion people, let alone the projected 9.5/10 billion by 2050, so an agriculture based on small mixed farms is a non-starter.

Clearly, they say, we need high-input industrial farms on the largest possible scale.

As is so often the case, the truth seems to be precisely opposite. On the small-scale, many a study shows that output of food from small mixed plots, when they are well run, far exceeds the output even of the most intensive industrial monocultures - mainly, it seems, because small plots are given plenty of tender loving care; the husbandry is intricate and high class. *Of course* farms of 1,000 hectares produce more food than farms of one or two hectares, but in a finite world what counts is the amount of food produced *per hectare*. What counts, too, is the amount of extraneous energy (oil, human or animal muscle power, and the rest) that must be expended per unit of food energy produced. Way back in 1962, in 'An Aspect of Indian Agriculture' *Economic Weekly*, the Nobel prize-winning economist Amartya Sen showed against all the expectations of the day that productivity per unit area is *inversely* related to the size of the farm. In other words, yield per hectare *increases* as farms grow smaller (other things being equal).

The same has now been shown over and over - in India, Pakistan, Nepal, Malaysia, Thailand, Java, the Philippines, Brazil, Colombia and Paraguay. Fatma Gül Ünal's study in Turkey in 2005 showed that farms of less than one hectare are 20 times more productive than farms of 10 hectares or more. In Britain there may be little doubt that yields from the best allotments and indeed from patios, area for area, may far out-yield even the most intensive industrial holdings.

More research is needed on this: it is one of the most significant observations of modern times with profound consequences for the world. We need to pin down all the details. What, for example, is meant by 'productivity' in any one set of circumstances? Biomass? Protein? Food energy? The protein and energy content of a kilo of wheat is far higher than that of the average fruit or vegetable. To what extent are we comparing like for like?

All the signs are that Sen's observation holds true for most of the world most of the time. This absolutely negates modern agricultural strategy, for the mantra from on high is that scale-up is vital: more and more inputs on a bigger and bigger scale. Already there are farms in the Ukraine of

300,000 hectares and more - bigger than most English counties. In the Cerrado of Brazil (the vast dry forest where Brasilia is sited) there are plantations of soya and sugar cane of hundreds of square kilometres - and each square km is 100 hectares. (The soya is intended to feed European livestock including cattle for whom soya is a seriously alien food. The sugar cane is for burning; grown not for food but for 'biofuel'.) *The Economist* is wont to refer to the Cerrado as a wasteland but it is in fact among the world's richest ecosystems, and a wondrous forest garden, among other things, producing hundreds of species (literally) of edible plants. But government ministers, industrialists and their chosen experts continue to insist with worried shakes of the head that *unless* we merge the world's farms to create as few as possible, and as big as possible, and replace as many farmers as possible with big machines and industrial chemistry, then we will all starve. This simply makes no sense - except in terms of commerce and power.

Even without more detailed research, the case for small mixed farms already seems to be open and shut. The IAASTD report cited in Chapter 1 pointed out that small farms currently produce 50 per cent of the world's food. Another 20 per cent of our food comes from fishing and hunting (though as things stand, both are horribly unsustainable) and people's back gardens. Thus the vaunted industrial agriculture that is deemed to be essential provides only 30 per cent of our food.

As many Third World observers have pointed out, most of the world's small farms are located in the poorer countries in Eastern Europe and developing countries, and typically produce far less than they could - not because they need more high tech or smarter farmers or more highly bred cows, but for logistic reasons. Thus in an Indian village a few years ago I asked why the farmers didn't produce more milk, which they could easily have done, to sell to the local town for cash. Because, they said, the cows are in milk when it rains, and when it rains you can't get down the road to the town. Worldwide, small farmers typically don't invest, what's needed to raise output because in the present economy prices are not guaranteed, and if they doubled their crop at twice the expense they could well find at the end of the season that the price on the world market is down by 20 per cent or more - enough to make nonsense of all their endeavours. Markets can fluctuate in days - the price of wheat in Britain in March 2014 jumped by £6 a tonne in instant response to the political upheaval in Ukraine (which of course became far worse); but farmers need at least one season to build up for the next crop and a serious change of direction can take a decade. The timescale of the market is out of step with the biological-climatic timescale of agriculture. Farmers who want to make a killing these days must follow the futures market, which (*vide* Ukraine and drought in Brazil and a few little inconveniences like that) is somewhat less reliable than crystal-gazing. Small farmers with marginal incomes can't afford to take such risks so they keep their inputs to a minimum, and output suffers accordingly.

Professor Bob Orskov of the James Hutton Institute in Aberdeen was born and brought up on a farm himself (in Denmark) and has spent many years working with small farmers the world over. He points out that small farmers in poor countries could generally double or triple their output - they have the know-how - if only the logistic problems, often crude and obvious, were ironed out. What they don't need are high-tech varieties of crops and the highly productive, super-milky Holstein cattle which the rich countries are only too anxious to supply them with, so as to balance their own books. (Saudi Arabia and Israel have established enormous herds of Holsteins in their various deserts which is biological nonsense of the highest order and a huge waste of fossil water and fuel.)

By contrast, high-tech crops and livestock in industrial countries are often already producing as much as seems biologically possible while animals are being pushed beyond what is morally acceptable. Thus the *average* yield of British dairy cows is now around 6,000 litres a year (1,200 gallons), which is about four times the output of a wild cow, while 'elite' cows from breeds like the Holstein produce 10,000 litres-plus (2,000 gallons). They have udders like dustbins which they must straddle; lameness is almost the norm. To maintain such output their metabolic rate is raised by about three times. The overall strain is such that they are lucky to get through three lactations, while a wild cow or one kept traditionally can often get through ten or more. But still the industrial agroscientists press for more.

None of this even makes sense - if, that is, it was truly the intention of leaders and politicians to feed the world. It would be difficult, even heroic, to raise the output of industrial farms by more than another, say, 20 per cent. Yet this would raise the total food output by only about 6 per cent. But if the output of the small farms in poor countries was doubled, which would be far easier, total output would be increased by more than 30 per cent. *If* we decide we still need more food, then we should be seeking to make life easier for the millions of small farmers in economically developing countries - not to replace those farms with high-tech estates as is the Western way. Yet most of the world's endeavour right now, driven by corporates but largely supported by taxpayers, is aimed at industrial agriculture. This misdirection seems more than stupid. It is downright wicked.

The second argument brought to bear by those who defend the status quo is that in our attempts to produce enough by the methods of agroecology we would inevitably finish up with a diet of lentils and muesli - altogether too boring and non-sustainable for social reasons. Again the truth is entirely opposite. As we saw in the last chapter, the kind of farming the world needs is just what's required for excellent nutrition and the world's finest cuisines.

Thirdly, say the detractors, food produced by skills-intensive methods would increase the cost of food hand over fist and many more would starve. This again requires more detailed discussion (see Chapter 13), but two points stand out. One: we should *not* reduce the price of food but we should reduce the price of everything else - all the other drains on our wealth that are imposed by the modern economy. Two: in countries like Britain and the US, food is sold mostly through supermarkets at the end of a long and tortuous food chain. This is nothing like so secure as we might hope, as was spectacularly revealed in 2013 when it transpired that some of Britain's best-known beefburgers contained horse. Much more to the point is that the food chain - the middle men - soaks up about 80 per cent of the total retail price of food. The farmers get only about 20 per cent at most. Ministers and experts continue to tell us that to increase efficiency we must cut farm labour, but farm labour in a modern industrial system can hardly account for more than 10 per cent of the total food cost (and farm workers are citizens too, and taxpayers). Clearly the way to keep food prices within bounds is not to sack farmers. We need shorter food chains, meaning different kinds of markets. More in Part III.

Fourthly, there's the spurious sociological argument: that the appeal for small mixed farms is simply an exercise in nostalgia and/or Luddism (as usually understood) - that enlightened agriculture is simply a flight from high technology. Not so. I am advocating small mixed farms not primarily because they are picturesque (although they typically are) but because, when not actively done down, they demonstrably work. They are what the world needs. In contrast, dedication to the industrial model is *pure* ideology: belief in the dogma of neoliberalism and faith in high tech. As for Luddism: the small mixed farms of the future need excellent science just as much as the industrialists, and can benefit as much from high tech. In structure the small mixed farms we need may resemble those of the past but in detail they are likely to be very different - technically, and in their social structure (as discussed in Part III). But the science that's needed in the future must be far more subtle than the kind we have now. Above all, we need the science of ecology, the most complex science of all, so we can work creatively with nature. The agricultural science we have now, though perceived to be ultra-modern and on the ball, is seriously old-fashioned. It is rooted in the long-discredited eighteenth-century conceit that we really can understand nature exhaustively and manipulate it at will, and in nineteenth-century gung-ho technology - notably in industrial chemistry. The much-vaunted GMOs (genetically modified organisms) that now soak up so much of the research budget are just bells and whistles, clever but ill-conceived and contributing nothing to our food security that is truly worthwhile. Neoliberal economics, too, which drives all modern strategy, is rooted in eighteenth- and early nineteenth-century ideas - those of Adam Smith and David Ricardo - that need serious refinement. Indeed, what now passes as modernity is just a veneer. Real modernity requires root-and-branch rethinking.

Finally, an influential gentleman from the London School of Economics once assured me that, since farming is such a terrible way to make a living, farmers should be grateful to be thrown out of work. Recently, an adviser to an international bank earnestly told me that life in the slums, which in poor countries is the final refuge before the grave for the many millions who are being thrown off the land, is really quite jolly. The people have a whale of a time recycling old tyres and polythene bags, and there is music and dancing almost every night. Such arguments are becoming fashionable in high places. Our leaders are more and more anxious to assure us all and perhaps themselves that, thanks to their ministrations, and despite appearances, all is well with the world. It's the same with climate change, say the powers that be. Don't try to keep it within bounds, for that would be too disruptive. Get used to it. These are the arguments of the desperate; of people who, if they have any insight at all, must know they have lost the plot and are looking for forms of words to justify the nonsense that their strategies have led us to.

If we acknowledge that the world really does need hands-on farmers - we should not try to replace them all with robots - and that farming really can be among the most rewarding jobs of all, we should ask (as very few governments have done, to my knowledge) how many of them we need. What is the proper balance between the agrarian and the urban economy?

## How many farmers does the world really need?

With profit-driven agriculture in a world where the oil still flows and big machines and industrial chemistry provide a technical option, the answer is obvious: the fewer the better. Though it is hard to pin down the statistics because there are so many part-timers and so much casual labour, it seems at present that in Britain and the US only around 1-2 per cent of the working population work full-time on the land. Farmers in Britain, especially small- and medium-sized dairy farmers, continue to go broke and disappear into the social ether. Thomas Jefferson saw the emerging US as 'a country of small farmers' but now the number of farming full-timers is roughly matched by the number in prison or on probation. In Rwanda, in sharpest contrast, 90 per cent of the workforce is on the land. In Angola it's around 80 per cent. In developing countries in general, including India in particular, it's 60 per cent. Britain and the US both see themselves as world leaders. Development, in practice, has largely meant following the Western lead - and particularly the US lead. It is often taken to be self-evident that those countries who seek truly to modernize and develop must cut their agrarian workforce. When Turkey was pressing to join the

EU, it was advised to prune its farmers from 30 per cent of the whole to 18 per cent, since agriculture at the time contributed 18 per cent of the GDP, which was taken to be the only worthwhile measure of value.

Is this the right approach? Well, it is easy to see that Rwanda's 90 per cent is too many. It means there are too few non-farmers to provide a home market, so without exports the farmers can't get past subsistence. Clearly, too, there are not enough non-farmers to supply all the doctors and teachers and bus drivers and builders and other key roles that a properly functioning society needs. For men in much of Africa the only alternative to subsistence farming is mercenary soldiering and for millions of women, who in practice do much or most of the farming, the options are even less promising. That doesn't seem right: not what the world ought to be aiming for.

So what is the right number of farmers? Many an agricultural economist insists that the fewer the better. Labour is expensive so sacking people reduces costs and this increases efficiency, when efficiency is measured in short-term cash, and we don't cost human misery. Yet Britain's 1-2 per cent farmers - which some would like to reduce still further! - is surely too few. Certainly, with so few farmers, enlightened skills-intensive agriculture is a non-starter. Then again, 60 per cent of Indians working on the land means 600 million people. If India farmed the industrial way most of them would be out of work. What should they do then? I was told in India that there are alternative industries - and of course there are: India's economy and its industrial infrastructure have been growing rapidly. Like China, it has been the world's sleeping giant.

But, as both the waking giants are beginning to demonstrate, no country can afford the labour-intensive heavy industries and factories that soaked up so much labour in Britain in the nineteenth century, and the Earth can no longer withstand the inevitable pollution from all the fuel they burn and all the toxic minerals they scatter through their mining. In practice, in India, alternative industries largely means IT and tourism. Indian IT is wonderful but it employs only tens of thousands - at least four orders of magnitude short of the hundreds of millions who would be out of work if India's farming was like Britain's. A job in tourism largely means cleaning hotels before the tourists get up, then a bus back to the shanty. For many millions in India, as in most of Asia, the end of the industrial line is the sweatshop, the modern equivalent of the Victorian workhouse. In a world that aspires to be humane and civilized that surely is not desirable. Sweatshops are the modern equivalent of slavery which the civilized world claims to abhor - except that traditional slave owners felt obliged to take care of their slaves (they were expensive) while sweatshop owners and employers of casual labour commonly feel no such obligation. In theory, the sweatshop workers are free agents, essentially sub-contractors, able to come and go - or that at least is the gloss.

In short, for most people in most countries, farming and all that goes with it are by far the best option. No other human pursuit could possibly supply all the jobs that humanity needs in a morally acceptable and sustainable fashion.

In the absence of any formal research, so far as I know, I would suggest as a top-of-the-head calculation based on conversations around the world, that no country, apart from those that have no farming at all to speak of like Monaco and Singapore, should have fewer than 10 per cent of the workforce on the land; and to keep the society balanced, probably none should have many more than 50 per cent. That, at least, is a good working range. So Rwanda should probably lose almost half its farming force (but only as jobs are created elsewhere!) while India is probably about right, at least for the foreseeable future.

Britain and the US, however, could do with many more farmers than we have now. My conversations with British farmers suggest that we could do with eight times as many: say a million more to kick things off. That is roughly commensurate with the number of unemployed under-25s in Britain right now. Educationally the unemployed are a very mixed bunch. But farming can make good use of all talents. That is one of its joys.

Needless to say, Britain has no plans even remotely on the horizon to train and recruit another million farmers. There are various apprentice schemes, but only at most for a few tens of thousands. Our leaders are wedded to minimum-to-zero labour farming geared to the neoliberal market economy and when the crops need picking we can always import a few busloads of Romanians and Chinese, at least till the end of the season, when they can be thrown out again. For the time being this works in cash terms - though not if we count the social cost, for there are few rougher deals than that facing the migrant worker. A government that presides over such a system is not a government at all in any worthwhile sense. But that is what we have. Clearly, if we want the world to be different we have to do what needs doing for ourselves. This is discussed in Part III.

What of the idea that farming in general is primitive, and fit only for big machines or for slaves? Again this is nonsense. To be sure, farming wouldn't suit everyone - and it clearly does not suit all those farmers' sons and daughters who can't wait to get away from it. There is no gene for farming. But then, it doesn't have to suit everyone. In Britain, one in ten would be enough. Farming should be offered as a serious career option for every child. Farming has to be a vocation, like medicine or teaching or playing the violin professionally. For those who love these things, they're the only thing there is. I always remember a doctor who said he felt sorry for people who have to do anything else, and farming has the same lure.

You see it in farmers of all ages who lean on the gate and just look at their sheep, for hours and hours; not idleness, but absorption. Yet for those who hate farming, but have no option, it is of course hell - just as it's hell for reluctant doctors and teachers. Many of today's farmers are miserable, and among farmers both in rich countries and poor the suicide rate is distressingly high: as Vandana Shiva relates in *The Vandana Shiva reader*, it runs to hundreds of thousands in India alone. But although I have often heard farmers moan, I have rarely, if ever, heard complaints about the job itself. Shepherds positively boast of February nights delivering lambs. They complain about the bureaucracy and the lack of reward; and in countries like Britain, with minimum-to-zero-labour farming, they are weighed down with the loneliness.

Third World farmers are depressed by their lack of kudos. Farmers keep us all alive but in many countries their status is rock bottom and for human beings, inveterately social creatures that we are, kudos is life-blood. In short, there is a great deal wrong with the job of farming right now but nothing that cannot be put right; and for those with an aptitude, it is the greatest job there is.

But there is one more key question that no government to my knowledge, and certainly no British government in memory, has properly addressed.

## Should we grow all our own food or import whatever is most economical?

The tension mentioned earlier - between the need to grow food for the home country, and the need to raise crops and livestock as commodities to sell to whoever will buy it to boost the income - should be resoluble, but is nonetheless real. In any one field we can either grow crops or raise livestock for home consumption or we can raise crops or animals for export. Which is better? Or - assuming both have their pros and cons - what is the right balance?

Except under very special circumstances, either option taken to extremes leads to absurdity, or at least to a less than optimal outcome. So if France, say, dedicated its fields entirely to wheat and vegetables and fruit (especially grapes) and livestock for home consumption, the people could surely eat wonderfully. But it would waste the opportunity to sell its fabulous produce - notably its wines - and so grow rich. On the other hand, over the past half-century in particular, and in line with the thinking of David Ricardo, many a poor country in the tropics has been persuaded in effect to abandon farming for the people at home and to concentrate on the crops that it can sell at a premium to foreigners - coffee, tea, bananas, pineapples, cut flowers, and all the rest. After all (the argument has it) the subsistence crops the exporting country would otherwise grow (sorghum, maize, cassava, whatever) are worth far less than the luxuries, and the US in particular has vast surpluses of basic crops such as maize which it is happy to sell off for a song, so it is far better to grow the expensive stuff and buy in the cheap stuff. The US can afford to sell its surpluses cut-price because, although it is the arch-defender of free enterprise and free trade, it nonetheless sees fit to subsidize its richer farmers with vast infusions of tax-payers' money: $77 billion on maize alone between 1995 and 2010.

Some countries at some stages of their history - including China, Japan and India for a time - more or less refused to trade at all with the outside world, especially in food; largely because they wanted, first, to get their own houses in order. Then, if and when they did want to trade with the outside world, they could do so from a position of strength. This strategy seems sensible enough, but it did come at a price, for their decades of isolation reduced their income and also reduced their food security because, without trade, it was difficult to provide enough food if and when their own crops failed. On the other hand, when former president Bill Clinton visited Haiti after the earthquake of 2010 and saw the people starving, he also saw how dangerous it was for such countries to abandon home growing all together in favour of commodity crops for export. Since his own government had encouraged such a strategy, he declared himself contrite.

The mists begin to clear, as such mists generally do, when we apply a little common sense and common humanity.

The two key concepts are those of food security - everyone should be sure they always have enough food; and food sovereignty - everyone or at least every country should have control over their own food supply. The two desiderata clearly overlap - but they are not the same. For instance, a person in a care home may have absolutely no control over his or her own food supply and yet feel perfectly secure, or at least as secure as anyone can be in this uncertain world, provided society at large knows of his or her plight and gives a damn. On the other hand, people may meet all the criteria of food sovereignty and yet be insecure - if, for example, they have cut all trade routes and then the crops fail.

On the whole, when everything is weighed up, the best course for all countries that can manage it is to balance self-reliance - implying food sovereignty - with fair trade: simple, commonsensical, and more or less unimprovable.

Self-reliance is *not* the same as self-sufficiency. Self-reliance means that a country should produce enough food to get by in bad times. Britain strove to become self-reliant during the wars with Napoleon, and then the two world wars, though it was always obliged to begin from a very low base because its governments never seemed to anticipate the need for self-reliance until blockade was upon them, and agriculture needs a long lead-time. *Most* countries could certainly achieve self-reliance. Britain certainly could if only we practised enlightened agriculture and farmed along agro-ecological lines. Several studies have shown this, but we can make the point with a few top-of-the-head statistics. Thus, the average yield of wheat in Britain is 8 tonnes per hectare. One kilogramme of wheat provides about 3,000 kilocalories plus commensurate protein, which is easily enough macronutrient for one person for a day. One kg per day means 365 kg per year which is about one third of a tonne (1 tonne = 1,000 kg). So one hectare at 8 tonnes provides enough food energy and protein to sustain 24 people. So Britain's 3 million ha (hectares) should be able to support 72 million people - as many as we are ever liable to have.

Of course, we can't live on wheat alone, but we also have another 15 million ha of agricultural land which is not arable. Much of those 15 million hectares could be arable, if needed, but at present it's grassland and woodland - ideal for livestock, with plenty of scope for horticulture. That's quite enough to complement the baseline of macronutrients provided by the arable. Britain, with its uplands and rainfall, is particularly grassy. What's true of Britain is true of most countries: in most, there is more grassland than dedicated arable. If the countries of Africa were in a position to farm closer to their full potential, then some of them could be self-reliant several times over, including some of those now perceived as disaster areas.

Self-sufficiency, on the other hand, implies that a country should strive to produce *all* its own food. So a self-sufficient Britain would be obliged to grow its own coffee and bananas - and although we can do this, under glass, it makes no sense to do so except for fun. It is better by far to buy from countries that can grow coffee and bananas easily, provided the trade is fair and sensible. Most obviously, the buyers should pay a fair price; and the exporting country should not sacrifice its own self-reliance so as to meet its trade commitments, as the Haitians (and many others) have been forced to do. Finally, the exporting country should not sacrifice its own biosphere, as is said to be happening in Kenya as the country grows more flowers and sugar-peas and other such fripperies for Western markets.

Much of the mess, the uncertainty, the malnourishment and the famine that is so evident in the present world comes about because these very basic, very commonsensical, and generally obvious principles are neglected. But then, it seems, modern governments and the rest of the oligarchs don't do common sense. They do power, and rhetoric, and misguided ideology.

All of this raises the multi-billion dollar question:

## Why don't we do the things that so obviously need doing?

The answer lies with the question posed at the start of chapter 1: should farming provide good food for all, without wrecking the planet, or maximize wealth? Under the prevailing neoliberal market economy, eked out by the creative accountancy of modern banking, the answer is clear: the maximization of wealth trumps all other considerations. The market is supposed to be free and fair, but a market that was absolutely free would lead us inexorably, in effect by the laws of physics, towards monopoly; and governments must intervene to prevent monopoly if only for their own benefit. But governments don't do more than they perceive they have to and in practice the free market is dominated by an oligopoly of corporates. So, for example, ten corporates control 80 per cent of the global agrochemical market; four control almost 100 per cent of the GM crop market (Syngenta, DuPont, Monsanto and Bayer); six processors control 93 per cent of the UK's dairy processing (including Dairy Crest and Nestlé); and five supermarkets control 80 per cent of the UK's retail trade in food (Tesco, ASDA, Sainsbury, Morrison's and the Co-op). There are still small farms and shops aplenty - SMEs - but governments prefer to deal with corporates precisely because there are so few, which makes them easier to deal with, but also, in practice, because they may offer ministers places on the board, at least after retirement, and it is good to have a choice.

If you want to maximize wealth - which in practice means maximizing profit, the difference between receipts and outgoings - then there are three things you have to do, whatever the nature of the business: and all of them are at odds with the requirements of enlightened agriculture.

First, you need to maximize turnover. As Jack Cohen, the founder of Tesco, commented about 100 years ago, 'Pile 'em high!' In agriculture, this means maximizing output. The principle of enough's enough goes out of the window straight away. Agribusiness has gone to huge lengths this past half-century and more to remove the ceiling on grain production so industrial farmers can sell more and more, first by feeding half of what we produce to livestock; then by burning it and calling it 'biofuel'. Governments like Britain's and America's present it as a virtue, although cereal-based biofuel is among the most flagrant of scams.

Secondly, to maximize profit you must add value: turn what is otherwise cheap into something more expensive before you sell it. To some extent adding value is good and necessary. Baking - turning raw wheat into bread - is good. Charcuterie is good - turning bits of pig ('everything bar the squeak') into high-value delectables is good. Brewing is good - barley and many other grains miraculously transformed into one of the world's pleasures and, indeed, into a valuable food.

But value-adding these days is all too often gratuitous - not intended to enhance our lives but simply to cash in, and often far more for the benefit of the retailer (as in additives to extend shelf-life) than for the consumer. Virtually all ready meals are in this category and bananas in plastic containers; and tasteless strawberries jetted in from Israel for Christmas. (Israeli growers told me in the 1970s, 'These strawberries are rubbish. But so long as you buy them we will grow them'. They do after all look good on the supermarket shelves. Dates, on the other hand, the traditional Christmas fruit, also from the Middle East, don't need jetting).

Thirdly, to maximize profit you must minimize costs - and this, in the context of agriculture, is the most pernicious exercise of all.

To begin with, it is extremely dangerous to try too hard to reduce the cost of raising livestock. Britain's livestock has suffered one epidemic after another over the last few decades. All were hideous. Some - mercifully not all - threatened humans too and people died. All were avoidable. All these epidemics were caused by bad strategy and/or false economy - careless imports, lack of infrastructure and cut-price husbandry.

First on the scene in the 1980s was BSE, bovine spongiform encephalopathy, or 'mad cow disease'. This was caused by feeding animal protein - bits of left-over farm animals - to cattle, especially dairy cattle, as a cheap source of protein, and then (to cut costs further) not preparing it properly. That was the *sole* cause. BSE does affect humans in the form of vCJD (variant Creutzfeldt-Jakob disease) and one leading epidemiologist in Britain estimated at one time that it might kill 100,000 people. (It hasn't done but the incubation is long in humans and we're not out of the wood yet). Thanks to heroic efforts, with enormous suffering among both animals and farmers (and a few more suicides), it has been brought under control. But it should never have happened. Normally BSE cannot affect cattle because cattle don't eat meat unless their feed is micky-finned with it. It's an anthropogenic disease (caused by humans), brought about by the perceived need to cut costs wherever possible.

Since then, we have seen swine fever, which affects only pigs; swine flu, which can affect people, and when it does is commonly lethal; fowl pest, aka Newcastle disease, which is confined to fowl; and a brush with bird flu, which again can kill people and often has in Asia. Apparently it was brought to Britain not by wild birds but by dodgy (but cheap!) imported chicken from affected areas.

Even worse, costs are reduced by cutting labour. This may seem to reduce the price of food in the short term (though this is largely illusory) but it also - of course! - puts people out of work. Unemployment worldwide is the royal road to poverty. In the world at large the erosion of farm labour in the spurious cause of cash efficiency must be the prime cause of the global poverty that governments tell us they are most keen to eliminate.

At least as bad is that without plenty of farmers on board the husbandry must be simplified. Subtlety gives way to heavy machinery and industrial chemistry and the essential qualities of polyculture and genetic diversity must make way for monoculture. Monocultures are extremely vulnerable to pests and diseases and must be protected with heavy doses of pesticide and herbicide. Monocultures exploit the soil nutrients less efficiently but output is maximized in the short term with heavy doses of fertilizer. The whole exercise is inefficient in terms of energy used and in biological terms, but is deemed nonetheless to be the most efficient because, with the price of oil and other inputs as they are, and with a marketing network geared to the supermarket, it is cash-efficient, meaning profitable.

This whole package - high-input, high-tech, low-to-zero-labour food production geared to the all-against-all global market - is the stuff of neoliberal-industrial (or NI) food production. Its advocates perceive it to be ultra-modern because industrial methods are deemed to be modern by definition, and the neoliberal market is seen these days to be the natural way of the world: the only kind of economy we can possibly contemplate.

In effect, NI agriculture - huge, high-input monocultures with minimal labour – is equated with progress. Technically, such farming is new - farm machines powered by fossil fuels have been with us only since the eighteenth century, and industrial chemistry since the nineteenth, and the neoliberal economy in its modern form dates only from the 1980s; a twinkling compared to the 10,000 years of recognizable agriculture. In short: NI agriculture is very much a novelty and still supplies only a third of the world's food, and yet is called conventional. Industrial farming is very clever technically and much of the science that lies behind it is invaluable. But teamed with the economic philosophy of neoliberalism it is deeply pernicious. It is flawed at every level: the intent of it, the details of husbandry, the economic and political framework, the science, the morality and indeed (a much neglected concept), the metaphysics. And everything else is compromised because of it.

Why should the world's leaders, the people who have most influence, fail to do the things that are so obvious and could serve us all well, and instead pursue a strategy that is so obviously detrimental? We will look at this in the next chapter.

CHAPTER FOUR

# Digging deep: an economy fit for farming

People in high places have often told me that my ideas are unrealistic. You can't make a living, they say, if you farm according to the principles of agroecology; or not unless you redefine agroecology, as some would like to do, so as to include technologies such as genetic engineering. Even if you can scratch a living you can't make as much as you could if you employed the industrial methods that are now called conventional - maximum output whether we need it or not, at the lowest cost, with minimum labour but plenty of high tech, and suitable tax breaks, all on the largest possible scale.

Classical economic theory argues that human beings are rational creatures who seek happiness, and that happiness grows with wealth. Therefore it is rational to want to make as much money as possible, and irrational to opt to do anything else. The fact that most people do not grow happier just because they are richer, at least once they are comfortably past the breadline, is not allowed to detract from the classical thesis. The fact that farming and other human activities that are focused on wealth to the exclusion of all else are demonstrably wrecking the whole world before our eyes, is not allowed to dent the conviction that the unswerving pursuit of wealth is rational.

So according to the oligarchs who run the world, to behave in ways that are threatening to kill us all off is 'rational', and to behave in ways that would allow our children and their children and other people's children and other species to thrive for thousands of years to come, is irrational. When madness has become the recommended course, and serious words have reversed their meaning, we really are in a mess. We need to take stock and start again.

I have already identified the approach that could ensure our food supply without wrecking the world: enlightened agriculture. But we cannot introduce enlightened agriculture in isolation. We need to prepare the ground: to create an economy that will support it, and install governments that will support the necessary economy, and indeed to adjust the entire zeitgeist: to ask ourselves what we really want out of life, and why. We - the whole human race - need to set ourselves a serious agenda, to rethink everything from first principles - as summarized in the following diagram:

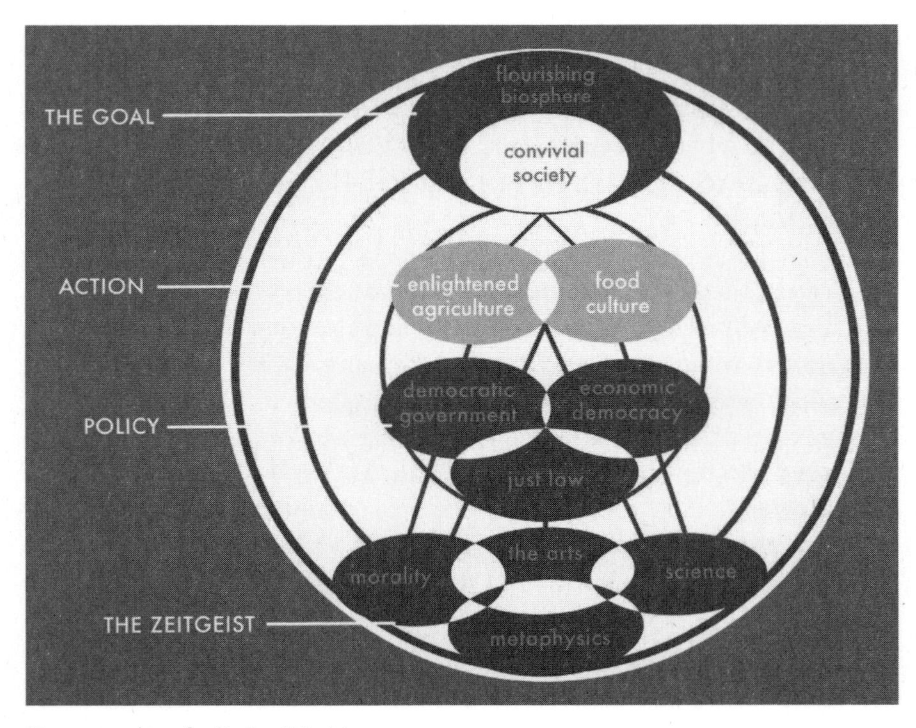


Components of a Better World

This diagram shows the things we need to think about, in a stack of eleven balloons. We need for practical purposes to think about them one by one, which is a kind of reductionism. But at the same time we need to think about each one in the light of all the others, which is one meaning of holism. Holistic thinking is vital, but is conspicuously lacking in high places.

As you can see, the balloons form four layers. The top layer is labelled 'convivial society' and 'flourishing biosphere'. Then comes 'enlightened agriculture' linked to 'food culture'; then the linked trial of 'economic democracy', 'democratic government' and 'just law'; and finally, 'morality', 'science' and 'the arts' all rooted in 'metaphysics'.

In a little more detail:

### *Convivial society in a secure and diverse biosphere*

The top two balloons define what I suggest we should be trying to achieve - which I am sure, deep down, most people would agree with: *convivial society* within *a flourishing biosphere.*

Convivial implies that we aspire to live together harmoniously, seeking to cooperate rather than to get ahead of the rest. We all need kudos, which means we all need respect; but we don't need to be at the top of the heap, which indeed is not, for most people, a very comfortable place to be. Ants manage to be cooperative without apparently thinking about it because, presumably, that's how they are programmed to behave. We human beings have a natural tendency to cooperate too, but we also think - and self-awareness gives us choice, as the book of Genesis tells us, and so we can choose either to cooperate or to bash each other's brains out for short-term advantage.

There are many reasons for choosing to cooperate: most of the time that's the easiest route, and there is always a chance of a quid pro quo. There is also the risk of retaliation: if I bash you and you nonetheless survive, you may come back at me with a piece of four-by-two. But that is not all there is to it. Mere enlightened self-interest does not provide a truly robust foundation for cooperativeness. We cooperate best when we truly care about the wellbeing of others; when we have compassion. This is where many people are inclined to balk. Such compassion is against human nature, they say. Nature in the end is 'red in tooth and claw', as Tennyson suggested. Life in general is a 'struggle for existence', as Darwin put it, and natural selection is driven by competition which - in Herbert Spencer's expression - leads to 'survival of the fittest'. We just can't afford to be too compassionate or our neighbour will take advantage.

As is so often case, the received truth is wide of the mark. As I argue in my most recent book, *Why Genes are Not Selfish and People are Nice*, and in line with many biologists, nature as a whole is *not* overwhelmingly competitive, and evolution is *not* driven exclusively by competition as many have inferred from Darwin's *On the Origin of Species*. When we survey nature across the board, we see that although competition is a fact of life, cooperativeness is its essence. For most creatures most of the time cooperation is the best survival tactic, so we would expect natural selection to favour attitudes that lead us to cooperate - which means we would expect natural selection to imbue us with compassion. So it has. Most of us know what it is to feel compassion and most of us prefer to cooperate than to fight. But, alas, we have allowed ourselves to be led by psychopaths - or, at least, by a philosophy that is fundamentally psychopathic. In truth to be *convivial* is the most rational course of all.

A 'flourishing biosphere' means an unpolluted world rich in species, all interacting within their respective ecosystems, each kind continuing its evolution in its own sweet way, forever. The word 'biosphere' should replace the conventional 'environment' - an anthropocentric term that literally means 'surroundings'. The term environment places human beings at the conceptual centre and treats all the rest as scenery - or indeed, which is what George W Bush meant by it, as real estate: or, more broadly, as a series of resources to be turned into commodities and hence into money, with all possible haste; for this is equated with progress.

The term biosphere was coined in the late nineteenth century by the Austrian geologist Eduard Suess to include the sum of all ecosystems, the whole living world; and extended in the early twentieth century by the Jesuit palaeontologist Teilhard de Chardin who emphasized that we, humanity, must feel ourselves to be part of it. Our concern for the biosphere must again be underpinned by compassion, extending to all species.

It is good to embrace the feeling that nature partakes of divinity. We need to nurture the realization that nature is not of our making and in the end is beyond our ken, and that each of us is privileged to briefly be a part of it. Science when properly construed encourages this feeling. The gung-ho brand of science that now prevails and claims to be hyper-modern in truth is an aberration of the eighteenth and nineteenth centuries.

Although the past few centuries seem to have favoured the idea that nature is ours to do what we like with, we can also discern a far more positive trend. Thus in the late eighteenth century the great Scottish geologist and theologian James Hutton opined that the Earth must operate in cycles - that, for example, the continued erosion of mountains must be balanced by a process of rebuilding. A hundred years later came Suess's idea of the biosphere, adopted by Teilhard. Then, beginning in the 1960s, the English scientist James Lovelock observed that the biosphere is homeostatic: it works in ways that perpetuate its own survival. In other words, the biosphere acts as if it were an organism in its own right. He called this conceptual organism 'Gaia'. The world has witnessed many great insights in the past half-century but none is more significant than this.

We do need the top two balloons to spell out what it is we want to achieve. Most modern governments and their commercial and academic confrères do not do this. Their only coherent ambition, it seems, is to achieve economic growth, apparently at all costs. Even at the cost of killing us all.

### *Enlightened agriculture and food culture*

The second layer of balloons down includes all the things we physically need to do in order to produce convivial society within a biosphere. If this was a general book, this layer would be labelled 'action', and might include building, medicine, social care, teaching, cooking and a great deal more. But agriculture and food culture may properly claim to be the most important endeavour of all, and we need farming and cooking that are expressly designed to nurture the biosphere and to foster conviviality, so that's what I've included here.

What enlightened agriculture entails is described in Chapter 1, and the concept is developed throughout the book. My point here is that farming both affects everything else and is affected by everything else, so it cannot be put to rights unless we attend to everything else as well. But most colleges of agriculture - or indeed, most modern centres of learning in all subjects - make little or no attempt to put the subjects that they teach into context, which is one reason why we have such a disjointed world.

The next three balloons describe the engine room of all societies: the mode of government, the economy, and law that is on the side of justice.

### *Economics and politics*

We cannot have enlightened agriculture or anything else on a secure basis, unless the economy is conducive to it, and the government supports the appropriate economy. Today, certainly in Britain, the economy and governance appear to be hostile to good farming practices and to conviviality in general.

I reckon there are two fundamental principles of governance, both of which now are flouted. First, government should be on the side of humanity and of the biosphere. British politicians like to claim that they are on our side but the fact that they need to make this claim explicitly suggests it is not the case. Most governments worldwide operate not as our servants but as our rulers, though in so-called 'democracies' like Britain's they are content to operate as part of a ruling oligarchy, with the corporates and banks setting the tone. Such governments support a predatory or a parasitic elite. Britain is now officially recognized as the least equitable society in Europe - the five richest families own as much as is owned by 12.5 million average citizens, and the gap between rich and poor continues to grow. We fall very far short of Abraham Lincoln's ideal: government of the people and for the people.

Secondly, the ideal system of governance is what neurologists call 'the neural net'. In the nervous and endocrine system there are control centres such as the brain and the pituitary and adrenal glands - but they are not dictators. They are subject to second-by-second (or millisecond by millisecond) feedback from the rest of the body (just as DNA is subject to constant feedback from the rest of the cell). Overall, we see consortia of neurones shifting from task to task, and reconvening for different purposes, with the leadership of the consortia constantly shifting. No system of governance works more effectively than the nervous-hormonal system of the body and if our political systems could emulate that we would surely be as close to the ideal as we could get.

Such an ideal clearly requires massive and concerted devolution of power - not just from central to regional and local government, but also to the many different interest groups in our societies. In Britain these should include bodies such as the National Trust or the Royal Society for the Protection of Birds, the religious bodies and all the professions (teachers, doctors, nurses) and the trades (we should bring back the guilds, in truth a more sophisticated concept than the trade union). Bodies like the Scottish Crofting Federation or La Via Campesina (the International Peasant Movement), should have real, recognized, statutory power.

To some extent, the EU encapsulates this idea in its principle of subsidiarity, which says that nothing should be governed at a higher level than could be achieved more satisfactorily at a lower level. Here is yet another enlightened principle that is decisively flouted.

### *An economy fit for farming: economic democracy*

A common and fundamental mistake is to suppose that the economy is simply about money, and that it is, or can be, morally neutral. In truth, it is the medium and the mechanism through which we translate our aspirations into action; our dreams into reality. These days the economy is conceived solely as a game of money, and those who conceive it that way don't see that this has enormous moral connotations, since it sidelines all other values like kindness, justice and aesthetics.

So behind agriculture we see two distinct schools of thought. The kind advocated in this book suggests that the point of farming is to provide everyone with good food without wrecking the rest of the biosphere - which I am calling enlightened agriculture. The second school of thought suggests that the point of farming - like everything else - must be to maximize wealth.

This latter way of thinking is formalized in the economic philosophy known as neoliberalism which first appeared in modern form in the 1960s and was adopted in the 1980s by Margaret Thatcher in Britain and then by Ronald Reagan in the US, and has now become the global norm. All the major political parties in Britain have subscribed to it, including Labour for whom it should be an anathema; and so, too, disastrously, does the all-powerful National Farmers' Union (all-powerful apart from the corporates, that is).

For obvious reasons the corporate-led money-driven market favours industrial farming which is expressly designed to benefit from it. But enlightened agriculture foolishly argues that life, human and otherwise, is more important than mere wealth, especially when that wealth remains in the hands of a small minority. Accordingly it is ignored by the powers-that-be and at every opportunity derided. Some who would like in principle to realize the ideals of enlightened agriculture - to create convivial society; to promote kindness to animals; to ensure a world with wild creatures in it - feel that the way to achieve this is by doing deals with corporates: accepting grants from big meat traders for research into animal welfare; producing food to enlightened standards to be sold in one or other of the big five supermarkets. Those who go down this route argue that if they can persuade big and powerful players to change their ways, ever so slightly, this will have a big knock-on effect across all of farming.

Some of the on-farm research backed by corporates is of great value, and many a small producer is able to make a living only through deals with Waitrose, say, and we all have to make a living. But although the compromise sounds sensible and plausible it is at best a stop-gap. The overall structure of the food chain that is geared to the neoliberal global market and demands maximum short-term profit is quite different from the structure needed to support enlightened agriculture. The neoliberal food chain demands centralization - big monocultures from all around the world delivering to huge depots that distribute to supermarkets; and in practice 80 per cent of what is spent on food in the supermarkets goes to the food chain itself, and to its shareholders. As outlined in Chapter 1, enlightened agriculture needs small mixed farms which require markets that are as local as possible - with 60 per cent plus of the retail price going to the farmers. That way they could afford the complexity and the workforce required to farm well *without* raising the retail price.

Here we encounter what should be only a minor glitch - but also a huge serendipity. The glitch is recent history compounded by ignorance, for a whole generation has now grown up which equates neoliberalism with capitalism. Some say that those who question neoliberalism must be anti-capitalist. People are encouraged to equate the 'free' in 'free market' with freedom in general so that anyone who questions the wisdom of the neo-liberal market must be a 'commie': some kind of Stalinist or Maoist advocating a centralized economy minutely controlled by government. The American Right argues this with enormous success - persuading people at large to compromise their own lives, mightily, so as to support the perceived right of the extremely rich to grow even richer.

But here's the serendipity. We do indeed need to change the economy radically but we do not need to 'smash capitalism', or to follow Lenin and Stalin, in order to achieve what's necessary. In fact, the economic mechanisms we need are all perfectly respectable - the kind that traditional British Tories and the traditional Labour Party would both have been happy with before neoliberalism swept all before it. All the mechanisms we need can be found beneath the broad capitalist umbrella. We can be sufficiently radical, achieve all that we need to, without being revolutionary in the Stalinist or the Maoist sense.

All we need, in fact, is **economic democracy**, which accepts (as centralized economies do not) that the basis of the economy is business, broadly defined - but business bounded by principles of morality. In contrast to some extreme Left positions, too, it accepts that money is a vital lubricant and a way of keeping track - but it is not, as in the present prevailing economy, the *point* of our endeavours. It also acknowledges that trade is vital. These are the basic premises of capitalism, broadly conceived.

Then we add five further refinements.

First, all businesses, of whatever kind and whoever owns them, should be conceived as **social enterprises**. This means that they must 'wash their faces' (ie not lose money) but their primary purpose is not to maximize profits at all costs but to bring about improvements in society and/or in the biosphere.

Secondly, we need the idea of the **tripartite mixed economy**, which I ascribe to Martin Large, and to his book *Common Wealth*. Ownership should be part private and part government, as in the traditional idea of the mixed economy which both of Britain's major political parties espoused until 1980; but also with a third component - **community ownership**. Communities can be defined topographically (village, neighbourhood) or by interest group, as in, say, Martin Large's Biodynamic Land Trust. There should, he says, be more and more emphasis on community ownership, and he is surely right. It should indeed become the lead player.

Thirdly, the private component of the tripartite economy should be dominated by **small and medium-sized enterprises** (SMEs) which remain under human control and are conceived as social enterprises. Small businesses with a social conscience - this is the kind of economy that in the radical 1960s we used to deride as *bourgeois* or *petit bourgeois*. But after 30 years of neoliberalism, *petit bourgeois* is emerging afresh as the great desideratum. Like 'Desert Rat' or 'suffragette', the erstwhile term of abuse is becoming a badge of honour.

Fourthly, both private and community-owned enterprises should be funded primarily through **ethical investment**, better called 'positive investment'. That is, people at large investing what they can afford in enterprises that they feel in their bones are worth supporting.

Finally - and this is a grand idea that encompasses all the rest - we need a **circular economy**. Present-day economies are linear. We start by mining some resource; then turn it into goods; use the goods; and then throw what's left away. Some of the resultant rubbish (but in practice usually only a small percentage) is recycled, but the general effect is a headlong plunge into entropy. We are running the world down.

In the circular economy, in absolute contrast, goods should be designed, in effect, to last forever. Complex machines should be assembled in such a way that they can readily be disassembled, and the more robust components used again, while the rest find other uses. Wherever possible, too, goods should be hired rather than owned. Then it is in the manufacturers' interests to ensure that their products are built to last - not, as now, to build in obsolescence, so that we have to buy replacements every few years. More and more companies and interest groups are taking the circular economy seriously. Between them, the circular economy and economic democracy offer real promise.

You may feel that the economy I am advocating here is simply that of the 'triple bottom line', proposed by John Elkington in his book *Cannibals with Forks: The triple bottom line of 21st century business*. This says, as I am, that the economy must be guided not simply by money but by a combination of social, environmental (or ecological) and financial considerations. But the triple bottom-line idea seems to place equal stress on all three pillars, and the economy is commonly taken in practice to mean the neoliberal market. In contrast, I am arguing that moral (social) and biological (ecological) ideals must be the drivers. The economy must be seen not as the third pillar, an aspiration in its own right, but simply as a pragmatic device that enables the primary ideals to be met. J M Keynes argued this, as many other economists have done.

The systems of governance and of economics may be seen as the engine room of society and both must be supported by law that does not favour the power-groups. The bottom three balloons - morality, metaphysics, science - deal with the underlying ideas and define the attitudes that form the zeitgeist, the spirit of the age, and inform everything that we do.

### *The zeitgeist*

The ideas and the attitudes that form the zeitgeist determine what we think is important; yet most of us, most of the time, hardly stop to examine these big ideas although they shape our lives. This is dangerous. It's because we don't bother to explore the notions that clutter the basement of our minds that we accept, for example, the Neo-Darwinian nonsense that lies behind neoliberalism and tells us that the natural way of the world is to compete, and that the natural course of humanity is to be at each other's throats. We need to drag these ideas from the depths and examine them. Detailed discussion belongs in another book which I haven't written yet (the sequel to *Why Genes are Not Selfish*), but here are a few salient observations.

I suggest that the zeitgeist has four principal components: morality, which tells us what's right and wrong, and how we *should* behave; science, which aspires to tell us what the material universe is like; and metaphysics which, says Seyyed Hossein Nasr, Professor of Islamic Studies at George Washington University, DC, asks 'the ultimate questions'. Finally, the arts are the joker: they shape our attitudes, our questions and our conceits.

Ever since the late eighteenth century, at least in secular Western circles, ethics (the term tends to be used synonymously with 'morality') has primarily been utilitarian: goodness or badness is judged purely by outcome. What's morally good, says the rationale, is what brings about 'the greatest happiness of the greatest number'. This may sound sensible enough, but it raises serious questions. If six thugs beat up one pensioner, and get a great kick out of it, that's six happy people and one miserable one - yet this can't be good. Worse: when the market is allowed to dominate as it does now, what is judged to be good is what's most cost-effective - which is why zero-labour industrial farming in a world where oil is still cheap is considered to be morally OK. Worse still: the market these days determines all morality, so what's considered good is what people will pay for. On the one occasion I attended the World Economic Forum in Davos, I heard human cloning defended on the grounds that there was a market for it.

In absolute contrast, more and more moralists agree that we need to return to 'virtue ethics', rooted in what most of us feel in our bones to be good. The three prime virtues that lie at the heart of all the great religions are compassion (the absolute essential); humility; and reverence for the biosphere.

All these notions are supported by truly modern science. For although truly modern evolutionary thinking is rooted in Darwin's ideas (he really was the greatest), more and more biologists agree that he overstressed the role and importance of competition. In truth, in nature we see a balance between competition and cooperation - but cooperation prevails, at every level. If it did not, life would not be possible at all. Natural selection, Darwin's great insight, favours cooperation. The neoliberal claim that their ultra-competitive, ruthless market is natural and therefore OK is bad science and even worse philosophy.

A truly modern philosophy of science should encourage humility, for it tells us that science can never tell us unequivocally what is true, never live up to the lawyers' standard: the whole truth and nothing but the truth. All its ideas are partial, uncertain and provisional. The universe itself is innately unpredictable - at the most fundamental level the uncertainty principle applies, and cause and effect is always non-linear - all causes having many possible effects, most of which cannot be anticipated. J S Mill pointed out in the mid-nineteenth century that however much we know, we can never be sure that we haven't missed something, and the history of science continues to reveal how much we do miss. In short, the confidence of modern science - which tells us, for example, that with genetic engineering we can tailor plants and animals to do anything we want them to and predict exactly what effect they will have on our health and on the biosphere - is the most terrible nonsense, and rather terrifying (as spelled out very comprehensively in Steven Druker's latest book, *Altered Genes and Twisted Truth*).

Such confidence did seem justified in the eighteenth century when science was still feeling its way. Nowadays we can see it as a serious anachronism. Agriculture does need excellent science: but the science it needs most is that of ecology, which is expressly designed to deal with uncertainty and non-linearity. Gung-ho science of the kind that has brought us the 10,000-litre cow and GMOs is perceived be ultra-modern but in truth, conceptually, it is seriously old-fashioned. To be sure, the details of science are dazzling. I find it impossible to renounce my subscription to *Nature*. But the philosophy that is conventionally brought to bear on science is crass in the extreme and the political and commercial excesses that such crassness leads us to are often so stupid as to make one gasp. What we now need more than anything, I reckon, is the confidence to see that the academic emperors we have allowed to dominate our thinking and our lives, have no clothes, or at least are seriously threadbare. They have lost their way and we have been seriously conned.

The central balloon, the root of all the rest, is metaphysics. As Professor Nasr says, we have allowed metaphysics to disappear from Western thinking, at least outside the formal religions, even though it asks the most fundamental questions; and, he says, this is humanity's greatest mistake of all. I take the fundamental questions to be: What is the world really like? What is good? How do we know what's true? And, 'How come?'

Hard-headed scientists claim that they alone can tell us what the universe is really like - but as the great twentieth-century zoologist Sir Peter Medawar observed, science in the end is, and can only ever be, 'the art of the soluble'. Scientists ask only those questions they think they are able to answer, or at least to address. In practice this means they consider the observable and measurable world - the material world and some of its manifestations, as in psychology. Science does not ask, because it is not equipped to do so, whether the material world is all there is, or whether there is more to the universe than meets the eye.

The idea that there *is* more to the universe than meets the eye, is that of transcendence, and the 'more' includes the idea of the animate universe - that intelligence, mind, may be present in all things; a property of the universe as a whole. From this derives the idea that there is an agenda behind the universe, a purpose. Such an idea is beyond the brief of science but it is perfectly rational (when rationality is defined properly) and there is no *a priori* reason to dismiss it. It is, I suggest, what most people believe deep down, which is why so many people worldwide align with some religion or other, and *most* claim to be spiritual. Spirituality implies a feeling of transcendence and this should make us at least open to the idea that nature as a whole may properly be seen to be animate and indeed divine. This feeling underpins the essential reverence for the biosphere just as compassion underpins the urge for conviviality.

Similarly, philosophers argue the principles of morality; and other philosophers ask how we know what is true (the discipline of epistemology). But the fundamental questions - what *is* morality? What *is* truth? - are in essence metaphysical. The final question, 'How come?' - how come things are as they are and the world works the way it does - belongs only to metaphysics. The great questions of metaphysics are not in the end answerable, but they certainly demand to be asked. Without the metaphysical underpinning, all our contemplations are rootless.

Finally, the zeitgeist is shaped very significantly by literature, the visual arts and music. George Eliot commented that fiction brings us closer to the truth than mere philosophy can. Picasso said that 'art is a lie that points to the truth'. Schubert and John Lennon have helped to define our attitude to nature and in the end, attitude is all.

But we should come back to earth. Part II addresses the central idea of enlightened agriculture.

PART TWO

# Agroecology

CHAPTER FIVE

# The absolute requirement: fertile soil

Agriculture depends absolutely on fertile soil; soil-less culture, aka hydroponics, can work very well but only on the small scale for specialist purposes. Indeed, in the words of Eve Balfour, founder of Britain's Soil Association in 1946, 'Take care of the soil and the soil will take care of the crops'. But in the heady circles of industrial agriculture, Eve Balfour's words have not been heeded. Soil has been seriously abused: as a spokesman for the UN's Food and Agriculture Organization (FAO) told a meeting in Berlin in April 2015, 'Globally, 33 per cent of agricultural soil - fully a third - is classified as degraded.'

By the same token, the science of soils has been seriously under funded this past half-century and at times has been positively done down: in the mid-1980s the UK government closed the publically owned and enviably equipped Letcombe Laboratory in Oxfordshire, the only one focused exclusively on soil research. The message seems to have been that soil is just dirt. Merely an anchorage point.

In practice, the fertility of soil depends both on the nutrients it contains (or doesn't) and on its intricate micro-architecture, which largely determines to what extent the nutrients are available to the roots; and the micro-architecture in turn is maintained by the trillions upon trillions of microbes of many thousands of different species that live in the soil, and interact with each other and with the roots of plants. All of these, plus the fungi and the host of small animals, form the soil biodiversity. But in 2013 the Joint

Research Centre of the European Union revealed that in 56 per cent of EU territory soil biodiversity is under threat. The JRC report also showed that soil is most threatened in states which have sought to block more effective soil protection. In this context, the JRC especially cited the UK.

For arable farming in particular, this past half-century has been conceived as a field exercise in industrial chemistry and heavy engineering, geared to the maximization of short-term wealth, at least for a few. All the subtlety, and all respect for the life of the soil, has been overridden. Nutrients are supplied from the bag; specifically, from the bags marked 'NPK' (nitrogen, phosphorus, potassium). Weeds are zapped with herbicide, and pests and diseases with insecticides, acaricides, molluscicides, fungicides, bacteriocides and every other -cide (where 'cide' derives from the Latin for 'killing'). The industrial fertilizers and the pesticides and the rest were mostly derived from the chemistry developed in the two world wars for making explosives and poisoning people, and the ads for agrochemicals in farming magazines still use the rhetoric of conflict - 'Wage war on pests'; 'Zap those weeds!', and so on. The essential microbes that make soil truly fertile have been caught in the crossfire. All that's left are various grades of clay and grit. Ploughed by increasingly heavy machinery (it's hard to find a small tractor these days), the clay is crushed into a pan, hard as ceramic (which in effect it is) which roots cannot penetrate. In recent decades (though less so now), the pans have been broken up by deep ploughing, by tractors as powerful as war machines. Meanwhile the surface dust blows away on the wind and washes away in the streams. The rivers run brown.

But sanity in the form of humility and sound biology could be making a comeback. Certainly, at least in Britain, more and more arable farmers are seriously alarmed by the state of their fields and by the ever-increasing costs; finding themselves spending more and more on oil-based chemicals for less and less return. Many are now seeking to regenerate soil structure and general stability by introducing sheep to provide permanent ground cover, protect what's left of their soil from rain and wind, and restore its organic content. In fact they are returning step by step to mixed farming, or at least are trying to (there's a lot of obstacles, financial and logistic, as outlined in Chapter 13). Obviously, this is a significant shift of farm practice. Even more than that, we could be witnessing what the American philosopher of science Thomas Kuhn in the 1960s called a 'paradigm shift' - a wholesale shift in worldview - from the age of industrial chemistry to the age of biology, and especially of ecology. Unfortunately, of course, the British and US governments and their chosen advisers have yet to catch up. In their minds, the agrochemistry industry is, as they said recently of the banks, too big to fail. In the short term the industry is seen as a prime driver of economic growth. The long term will just have to take care of itself. Doubtless our descendants will think of something.

So let's look at the soil itself - what it is, and what can be done to put things right. Simply by putting a soil sample in a jar and shaking it up to separate it out we see immediately that soil in its natural state has two kinds of components: mineral, and organic. We should look at each in turn.

## The mineral component

The mineral element is the terra firma: rock of many kinds ground down to various degrees by wind and rain and frosts, and by seas that have long since receded. The smallest particles form clay; larger particles are silt; and the biggest - typically one mm-plus - are sand. All three are silicon-based.

Soil that is heavy with clay is very dense, and clings tightly to water, so it becomes like glue when wet and is the stuff of bricks (literally) when dry. Wet clay soil is also airless, but roots need air. Wet clay soils are slow to warm up in spring so crops may be slow to get going. In olden days (so I was told in my *Farmers Weekly* days), farmers used to judge the soil temperature, and whether it was time to sow, by sitting on it, preferably after removing their trousers so as to get the full benefit for long enough for the cold (or incipient warmth) to make itself felt; but nowadays they have all kinds of temperature gauges, and indeed the temperature can be read by satellites in space. In any case, cold spring soils can be given a helping hand with plastic sheeting in various forms (high tech that in this context is appropriate) and/or with mulch. Sandy soils are more airy and warm up more quickly and crops like carrots prefer them, but if they are too sandy they can hardly hold water at all and plants that are not xerophytes (specialized for drought, like succulents) quickly wilt when the sun comes out. Most plants like a nice loam, with clay, silt and sand, in proportions that can ensure a good balance of water and air.

Soils may also contain a greater or lesser component of chalk, which is calcium-based. Chalk is sedimentary, originally formed in the sea mostly by planktonic protozoa called foramens. Limestone, chemically the same as chalk, is commonly added to reduce acidity and to help to break up clay.

## The organic component

To the chemist, organic simply means 'containing carbon'. Organic-rich soils, therefore, are carbon rich. In practice, the organic component includes the miscellaneous residues of decaying, once-living material, from animals, plants, fungi and microbes; plus the myriad creatures of all kinds that are still alive. That all this decaying and living stuff really is important to the structure of the soil and the wellbeing of the plants that grow in it, and of the animals that feed on the plants, was first revealed formally to the Western world by the English botanist Sir Albert Howard who, in the decades before and after World War I introduced the world to the arts of composting, which he learnt from native farmers in Indore, India. Eve Balfour picked up his baton: he died in 1947, the year after she established the Soil Association. Modern science has reinforced their message abundantly - showing that the ecology of the organisms that live in soil and to a large extent create it, is as complex as a tropical rainforest. These insights should be counted among the most outstanding of twentieth-century agricultural science - far outranking in importance the invention, say, of DDT, or indeed of genetic engineering, which won their initiators Nobel Prizes. The insights of traditional soil chemistry are still of prime importance, but now they are subsumed within a far broader understanding of soil.

Organic material is often seen as the prime source of plant nutrients and in nature it certainly is. But although organic material is always rich in carbon (by definition), it is not necessarily rich in nitrogen or phosphorus or any other of the standard plant nutrients. Peat and bark, for instance, are very low in nitrogen - but as we will see, the carbon content is vital anyway. In practice, the organic proportion of soil may range from near zero (in some arable fields) to virtually 100 per cent (in peat bog). Plants won't grow in pure peat because it lacks essential nutrients and no one that I know would attempt to grow wheat, say, on pure compost, even though compost should be nutrient-rich. But aside from such extremes it seems more or less impossible ever to have too much organic material in the soil. *Most* agricultural soils worldwide could do with a great deal more organic content than they have now, and many are definitely deficient.

The principal component of organic material - carbon - makes all kinds of contributions.

## Soil carbon

First, carbon in the form of cellulose (as in peat) provides the sponginess which is so vital - enabling soil to be both moist and airy, and easy to work. Indeed, the ability of organic material of any kind to regulate soil water is often its most important contribution. Thus Bob Orskov of the James Hutton Institute tells a tale of agroforestry in the Philippines, when dairy cattle were grazed in the coconut plantations. Critics complained that the yield from the cows was not impressively high. But Professor Orskov pointed up to the trees: '*There's* your harvest!' he said. For the dung from the cattle improved the soil texture so that it held more water - just the right amount - and the palms responded with far bigger crops.

We could surely, too, reduce atmospheric $CO_2$ very considerably, and hence mitigate global warming, by sequestering more carbon in the soil in organic form. For the world's topsoil contains about twice as much carbon as the atmosphere. It seems to follow that a 10 per cent rise in total soil carbon would reduce atmospheric carbon by 20 per cent - and that would reduce the present atmospheric load from 400 parts per million (ppm) of $CO_2$ to 360 parts per million, and so put us back in the safe zone. Most agricultural soils worldwide could easily take another 10 per cent of carbon and some of the more depleted kinds (not least in over-worked arable fields) could surely take another 60 per cent or more. These figures are speculative - there aren't enough serious studies to offer a proper analysis. It is clear, though, that through all the world's discussions of climate change these past 40 years, the possible role of soil carbon in reducing atmospheric $CO_2$ has been much neglected. Yet farmers could surely help significantly to mitigate global warming by adding more organic material to their soil (and they should be rewarded for doing so).

Carbon also plays various roles - though the research is still at a very early stage - in plant nutrition and plant health. Plants are commonly assumed to be exclusive *autotrophs*: they acquire their nutrients, the stuff of which they are made, in inorganic form (carbon as carbon dioxide, nitrogen as nitrate and so on) and their energy from the sun. Only animals and fungi are supposed to be *heterotrophs*: they require the carbon component of their diet to be supplied ready-made, in organic form, although fungi also derive many of the nutrients they need in inorganic form. Fungi tend either to be parasites or - far more commonly - saprobes: organisms that live on dead organic material, which is a specialist form of heterotrophy.

Yet plants are not exclusive autotrophs. Some are heterotrophic - including parasitic plants like dodder or semi-parasites like *Striga*, which is such a menace in hot countries (not least on maize), and the world's many insectivorous plants like the sundew and the Venus fly-trap; and a great many plants with sticky and/or hairy leaves may derive some nitrogen from the insects that are trapped on them even though they are not specialist insectivores. It strikes me, though, that the many plants that rely on mycorrhizae (symbiotic relationships between fungi and plant roots - more later) must derive at least some organic material from the hyphae (the fine, branching tubes that make up the body of a fungus) that pervade their roots. The flow of organic materials cannot simply be from plant to fungus. So we should concede that most plants are to some extent heterotrophic and that this could sometimes be of critical significance.

Why not take this notion one step further and suggest that *all* plants have at least some ability to absorb some carbon through their roots, in organic form? They surely would not derive significant amounts of energy from this source, but they could take in a great many organic materials of the kind that might in a human or animal context be called tonics, or cryptonutrients, as outlined in Chapter 2: recondite and complex organic materials that could have a very significant impact on plant health. Thus it is often claimed that plants grown in organic-rich soils often do better than those grown in soils that have little or no organic content and yet contain the same amounts of measurable nutrients (nitrogen, phosphorus, potassium, etc). It's as if the organic soils provided some factor X - a suggestion that scientists of the hard-nosed type may scorn as 'muck 'n' mystery'.

But do cryptonutrients account for the factor X that low-organic soils seem to lack? Do they account for the general healthiness often observed in organically grown plants? Again, this is surely worth looking at. If this turns out to be true then we would hope and expect that whoever eats the organically grown plants, including us, would also benefit. Such effects would probably be long-term and would not show up in the customary short-term trials (largely on rats) that are used to test the value of foodstuffs, or the possible dangers, and this would cause some scientists to reject the whole idea for lack of evidence. But in truth this would merely show yet again that the standard tests that we increasingly rely upon to direct our lives are inadequate, and that some scientists are all too ready to believe whatever is commercially expedient. We shouldn't give up on promising ideas to early. Sometimes evidence is elusive.

However - a big however - we would not expect cryptonutrients to be present in the soil unless the soil was rich, not simply in organic material but specifically in soil microbes, because they are the soil's principal chemists, the creators of those recondite molecules.

All this, too, is largely speculative. What is not speculation is that organic farmers and growers generally require and expect organic material to supply the standard recognized nutrients - certainly nitrogen and preferably phosphorus too - as well as improving soil texture with carbon. So:

## The two biggies: nitrogen and phosphorus

The N and the P in NPK are nitrogen and phosphorus - both of which plants require in substantial amounts. Nitrogen is a major component both of proteins and of nucleic acids (DNA and the RNAs) while phosphorus, generally in the form of phosphate, is a key ingredient of nucleic acids and of cell membranes, and has many other roles in cell metabolism.

There is no shortage of nitrogen, and it is easily recycled. The atmosphere contains around 4 billion billion tonnes of it - 79 per cent of the whole. But atmospheric nitrogen is in molecular, gaseous form, $N_2$, and as such it's of no use to plants. Plants need their nitrogen to be 'fixed' - converted into soluble ions - before they can absorb it, mainly through their roots, and make use of it. 'Fixing' in practice means reducing nitrogen gas to ammonia, $NH_3$ (or its ionic form, $NH_4^+$) which may then be oxidized to make various oxides of nitrogen known collectively as $NO_x$, which in turn are ionized (and further oxidized) to produce nitrate ($NO_3^+$). The main problem for the farmer is to supply the right amounts of nitrogen to the crops at the times they need it most - but not to supply too much. Surpluses tend either to sublimate - waft away into the atmosphere, not least in the form of nitrous oxide, $N_2O$, which is a potent greenhouse gas; or may leach into the groundwater in the form of nitrate, $NO_3^-$, which may further be modified chemically to become a dangerous pollutant. Ways of supplying the that's needed without these side-effects are discussed later, and in Chapters 9 and 10.

Phosphorus is also abundant - according to the *Encyclopaedia Britannica* it's the twelfth most abundant element in the Earth's crust, accounting for one thousandth of its weight - and it, too, should be easy to recycle. As a chemical element, too, it is more or less indestructible (unlike, say, oil, which is destroyed as it is used). So until the next cosmic catastrophe, billions of years in the future, there will never be appreciably less phosphorus in the world than there is now. Yet, paradoxically, agriculturalists worry about a possible shortage of usable phosphorus almost as much as they worry about the dwindling supply of freshwater.

For the problem is not the amount, but availability. Farmers apply phosphorus in the form of phosphate which they obtained traditionally largely from guano (bird poo) and more recently, since that is an obviously limited resource, from reserves of phosphate rock which are mainly in the Western Sahara, Morocco, China and the US, with Morocco being the chief exporter. How long the reserves will last is not agreed; but many authorities suggest that the world will experience serious shortages of phosphate within 100 years.

So what's to be done? The obvious strategy, which all countries could and should initiate immediately and could have begun many decades ago, is to recycle. The modern industrial form of agriculture works on a linear economy. New ingredients, including phosphorus, are fed in at one end; crops and livestock are eaten; and whatever is left tends to be treated as waste - often recycled, to be sure, but largely just disposed of. In particular, most of the food produced by farming is ultimately eaten by humans, and in much of the world human excreta, rich in precious phosphorus, is simply despatched into the rivers and out to sea. Phosphorus is naturally recycled back on the land by various natural processes: notably, it is taken up by microbes and then by various plankton and then by fish and then by sea birds which deposit at least some of it ashore as guano; and salmon, returning to the north Pacific coast of America, are eaten by bears and wolves whose dung is said to provide the redwood forests with a third of their phosphate needs. Farmers need more than these natural processes can provide and, although it is perfectly possible to extract phosphorus from the sea by artificial means, this is hugely expensive.

So what's to be done? Well, the world and particularly the Western world needs to change its attitude to human poo and wee. On farm and surprisingly acceptable recycling is offered by the traditional reed bed; fast-growing reeds and the huge variety of creatures that flourish among them extract the nutrients from sewage wonderfully. For cities there are high-tech extraction techniques to take out phosphorus. Here (in contrast say to GMOs with all the hype that goes with them) is another very valuable task for high tech that would truly be of service to humankind.

## The practicalities of nitrogen fertilizer

Soluble forms of nitrogen get into the soil in various ways. Quite a lot may arrive as pollutants, in soil or air, much of it emanating from agricultural fields. A surprising amount is supplied by lightning, which converts $N_2$ to $NH_4^+$. Much is supplied by nitrogen fixation in the soil, courtesy of soil microbes, of which more later. A lot comes from the decaying bodies of animals, plants, fungi and microbes, and/or is contained in animal urine and dung. Commonly, and usually preferably, dead material and dung are composted before adding to the soil: the material is converted by saprobic fungi and micro-organisms into a reasonably homogenous, fairly stable state. In industrial agriculture, much or most of the N, typically in the form of ammonium nitrate, $NH_4NO_3$, is fixed in factories by the Haber-Bosch process which combines atmospheric nitrogen ($N_2$) with hydrogen ($H_2$) to make ammonia ($NH_3$). This is then converted into usable forms and delivered in bags. If the necessary energy is supplied by renewable means, for example by solar-generated electricity, then Haber-Bosch is sustainable. In past times, nitrate was often supplied in part from nitrate-rich rocks and may still be applied as guano or fishmeal or seaweed. But if more than a little is used, or at least unless we take far more care than we do, then these natural sources are unsustainable.

For organic farmers, there are three prime sources of nitrogen. The greatest and most sustainable is the fixation of atmospheric nitrogen gas by microbes. Next in importance are green manures - green plants rich in protein and hence in nitrogen that are ploughed or dug into the soil or left as mulch; and, thirdly, the excrements of farm animals - preferably in the form of manure, especially composted manure, rather than as slurry. Many farmers in Asia in particular make extensive use of human manure as a source of nitrogen and, with phosphorus, and Westerners should make much more use of it than they do.

In the end, though, the key to organic farming (and indeed to all earthly life) is nitrogen fixation by micro-organisms. We should look at this more closely.

## Nitrogen fixation by microbes and from dead stuff

The creatures colloquially known as microbes are properly called *prokaryotes*. They are single-celled (although some of them sometimes form aggregates) and they lack distinct nuclei: they keep their DNA in various forms of packaging around the cell cytoplasm. The organisms that do have distinct nuclei are called *eukaryotes* and include animals, plants, fungi, seaweeds and the miscellaneous host of one-celled creatures sometimes called protists, including *Amoeba* and diatoms and many more. Protists are often too small to see with the naked eye but nonetheless with their complicated cells they are far bigger than prokaryotes.

Until the 1970s microbiologists assumed that all prokaryotes - all microbes - belonged to one great group called bacteria. But then in the late 1970s the American biologist Carl Woese, little known to the world at large although he was one of the greatest biologists of the twentieth century, showed that microbes include two quite distinct kinds of organism: both prokaryotes, but very different in their biochemistry and in the nature of their RNAs. One such group is indeed the Bacteria. The other one Professor Woese called the Archaea, known colloquially as the archaeans (though he tended to call them 'archaes', pronounced 'arkies'). These two groups are so distinct that Woese suggested they should be seen as distinct *domains* of life. The eukaryotes, with their clearly defined nuclei, then form a third domain. Now this 'three-domain' view of life is widely accepted. It is astonishing, though, that an idea so fundamental should have emerged so recently.

Equally salutary, is how little we know of the details of microbial life - at least of the majority that live in the world at large, as opposed to the tiny minority that are studied by doctors because they cause disease, or by industrial microbiologists because they are of commercial significance, like the lactobacilli that produce yoghurt and cheese from milk. Indeed, Professor Norm Pace, now at the University of Colorado, suggests that we have so far identified only about one in 10,000 of the different species of microbe that live in the soil. We have very little idea of what is really out there in the microbial world, or what they all do, and how they interact with each other and with the eukaryotes, including plants and animals.

Doctors are now beginning to realize the huge and various roles played by our own gut bacteria in nutrition, and in our general physiology and well-being (and this is now a prime topic in the leading scientific journal, *Nature*). We know least of all about the archaeans. When they were first recognized as a distinct group, biologists assumed that they lived only in recondite niches such as the Dead Sea, where little else will live. Now it is clear that they are ubiquitous, and are among the commonest organisms on Earth. But although they are such huge players in the world's ecology, conventional agriculturalists including industrial agriculturalists have largely overlooked them. Agricultural science these past 200 years has focused not primarily on microbiology but on chemistry.

It has been recognized, at least since the nineteenth century, that a variety of microbes fix atmospheric nitrogen: turning it into soluble ions that are a principal plant food. In effect they do what the Haber-Bosch process does, but at low temperatures (biology is far more subtle than industrial chemistry). Many N-fixing bacteria and archaeans live free in the soil constantly enriching it. Many plants, including crops such as maize, exude organic compounds from their roots and so create a zone that is particularly attractive to N-fixing microbes. Many other plants have gone one step further and provide havens for N-fixing microbes within their own structures. Thus, in the flooded rice fields of Asia lives a floating fern known as *Azolla*; and within the spongy texture of the *Azolla* leaves lives the microbe *Anabaena*.

*Anabaena* is a cyanobacterium - a bacterium that practises photosynthesis. But it also fixes nitrogen. Indeed, in traditional rice paddy fields, the *Azolla/Anaebaena* partnership is the prime source of nitrogen fertility - all supplied by nature, gratis.

Many plants provide special nodules in their roots to harbour N-fixing bacteria. Alder trees are among a suite of plants that keep colonies of N-fixing *Frankia* in their root nodules, which enable alders to live in dank places that are seriously nitrogen deficient.

Most important of all, not just to farmers but to all terrestrial life, are the plants of the family Fabaceae, which harbour bacteria of the genus *Rhizobium* in copious pink nodules on their roots. The Fabaceae was previously called Leguminosae, and its many members are still known as legumes. Two groups are especially important for farmers, and hence for humanity. The grain legumes, also known as pulses, include peas, beans and lentils; and many leafy legumes are grown for protein-rich fodder and as green manure, including clovers, lucerne (aka alfalfa) and vetches. Gorse, typically found on nutrient-poor, sandy-heathy soils, is a legume.

In the tropics, leguminous trees are key players, including the hundreds of species of acacia of savannah and semi-desert. Acacias are important to tropical farmers as fodder for all kinds of livestock, and various leguminous species are grown in tree and coffee plantations, both to provide the shade that those plants prefer and to enrich the soil. Leguminous trees of many kinds are among the principal species of tropical forests. There, and in all terrestrial habitats worldwide, legumes in concert with rhizobia feed themselves from the air and all the plants around them, too, which in turn support the whole trophic pyramid of animals. The phenomenon of biological nitrogen fixation, and the ubiquitous legumes in their thousands of forms, seem to me to reinforce the idea of Gaia; that the whole biosphere has evolved to behave collectively as a self-sustaining organism.

But also, microbes and fungi, often working together, are the world's principal saprobes, breaking down dead organic material and releasing the nutrients it contains. Nitrogen in whole organisms, whether dead or alive, is contained in proteins and nucleic acids, which in their organic form seem to be of no use as plant food; but microbes in the soil obligingly break them down to nitrates, which the plants can use (and to ammonia, which in unfavourable circumstances will sublimate away into the atmosphere, turning a potential resource into a pollutant). The release of nitrate is relatively slow, which has advantages and disadvantages. The advantage is that a soil well plied with nitrogen-rich compost will continue to release nitrate throughout the growing season so the crops need never be deficient.

The bad news is that crops in spring, when crops should be growing fastest and out-pacing the weeds, may find themselves short because the nitrates are not released fast enough to keep them topped up. I reckon (heretically) that there is a strong case here (one of several) for bending the organic rules, and adding a little industrial $NH_4NO_3$ produced by the Haber-Bosch process. This would not of course happen in nature and in that sense is unnatural but then, *all* farming is in a sense unnatural, and crops in spring are always expected to grow faster than wild plants would. Nitrate produced industrially seems to be identical to nitrate produced by microbes, so why not add it from a bag for the judicious quick fix? One objection is that industrial fertilizers are produced by mega-corporates whose influence on world agriculture seems disproportionate, and it is against the vital principle of food sovereignty to rely on them. The Haber-Bosch process also uses a lot of energy which at present is oil-based, and so seems unsustainable. But fertilizer factories could be community-owned and perhaps could be small, and the necessary energy in principle could be solar (energy is energy) - so why not? In enlightened agriculture, organic farming is the default position rather than the absolute: what should be done as a matter of course unless there is very good reason to do something else. Here I reckon there may sometimes be good reason to do something else. In general, the agroecologist should not reject industrial technologies simply because they are industrial. The point always is to be most *appropriate* to the task in hand; and the task in spring, when every day matters, is to get the crops away to a good start.

Key, too, in manures (excrement in portable form) and composts (organic material that may or may not include manure and has decayed into a new and semi-stable form) is the ratio of carbon to nitrogen: the C:N ratio. Organic material that contains a great deal of C but little or no N (like straw or peat) may be good for soil texture but fail to nourish the crops. Organic material with very little C and a great deal of N, like slurry or chicken manure, may provide far more N than the plants can deal with, and the surplus runs off (leaches) or disappears on the wind (sublimates); and may adversely affect the crops in other ways, for example by affecting soil pH. Organic farmers and growers variously recommend a C:N ratio of between 20:1 and 6:1. In general, the N content can be enhanced by adding fresh green material or manure to the compost, while C is raised by adding cellulose, including straw from bedding (which also contains manure) or indeed, on the garden scale, shredded paper. Compost making is really cookery. There is a lot of science in it, but in the end it's a craft. Yet the distinction between cookery, or gardening, and science is far from clear.

Some research scientists have green fingers and their experiments work. Others, however clever, don't, and it's better if they stick to thinking. (I once interviewed a scientist in New York in the early days of PCR - the polymerase chain reaction - the technique used in forensics to analyse very small samples of DNA. He had pictures of the Pope and the Queen stuck to the side of his extremely fancy machine.
'Why?' I asked.
'In this game,' he replied, 'you need all the help you can get.')

Anyway, the roles of soil microbes as fixers of nitrogen and as saprobes, agents of decay, are central. Organic farming may primarily be seen as a field exercise in microbiology. It is infinitely subtle and too complex ever to be analysed in exhaustive detail, which is why cynics are wont to write it off. Industrial agriculture by contrast is essentially an attempt to reduce biology to chemistry. Industrial chemistry delivers in the short term but overall, and in the long-term, it is too crude by half. For long-term survival we really must now graduate to the Age of Biology: primarily an age of ecology, abetted by physiology and microbiology. Agroecology, which mainly but not entirely means organic farming, is showing the way.

Finally, the many recondite organic molecules that microbes produce also influence, profoundly, the structure of the soil itself. The complex carbohydrates (sugar-like compounds) that they and the roots of plants produce help to break down clay and create the crumb structure, with its balance of water and air and easily available nutrient, which all crops prefer. Even more impressively, the pioneer soil biologist Elaine Ingham, who among other things is chief scientist of America's Rodale Institute, told the Oxford Real Farming Conference in January 2015 that the crippling soil pans created by the crushing wheels of big farm vehicles can be broken up effectively but gently just by adding organic material to the soil above. The microbes in the compost break down the clay and allow plant roots and worms to penetrate.

Microbes are not the only significant players in the biology of the soil. Vital too are soil fungi, both in their role as saprobes and in the partnerships they form with plants, known as mycorrhizae (the symbiotic associations that form between the roots of most plant species and fungi); yet another biological phenomenon that for a long time was largely overlooked, but can now be seen to be key.

## Mycorrhizae

Mycorrhizae, like the N-fixing bacteria in the roots of legumes and others, are a supreme example of *mutualism*: the form of symbiosis in which both parties clearly benefit. In a mycorrhiza, hyphae from soil-living fungi extend into the roots of plants, sometimes actually entering the cells of the roots, in some cases content to remain in the spaces between the cells. In both cases the whole mycelium of the fungus (the total collection of hyphae) becomes in effect an extension of the roots, hugely increasing their range and helping to gather all kinds of nutrients - notably phosphorus - that they might otherwise find difficult to prise loose. Possibly, too (I would say probably) the mycorrhizal symbionts provide small quantities of organic molecules, cryptonutrients and (probably) antibiotics, which help to keep the plant in good heart and to ward off pathogens. In reply the plant provides sugars, created in the leaves by photosynthesis. The plant-mycorrhiza consortium is like a lichen on a giant scale, combining the autotrophy of the plant with the heterotrophy of the cooperating fungus. Many thousands of different mycorrhizal fungi are known. Most wild plants make use of them - trees in particular, especially pines and their relatives which could not live without them; and some individual trees harbour several or many mycorrhizal species together. The fruiting bodies of mycorrhizae manifest as toadstools. In short, mycorrhizae are all but ubiquitous. Indeed it is possible that plants could never have come on to land unless they had been partnered by fungi in one way or another.

Knowledge of mycorrhizae is still scant, relative to their complexity and significance; and their importance in agriculture has only recently begun to be appreciated. Notably, it's become clear that perennial plants build up substantial mycorrhizae, while annuals may have little time to do so. It's clear, too, that the plant-mycorrhizal relationship is interrupted by cultivation. After ploughing, the crops and fungi more or less have to start again. Some plants - in general those that succeed in freshly broken ground - have evolved to live without mycorrhizae altogether. This is true of some crops, particularly annual crops, and of many of the plants that we call weeds. They achieve the status of weeds, and compete with crops, largely because they can do well in disturbed ground, without mycorrhizae, and can tolerate or thrive on soils that are far more fertile (richer in nitrogen) than is common in the wild. Most of the plants we commonly call wildflowers grow happily at the bases of hedges or in woods or permanent pasture year after year, dependant on their mycorrhizae, but cannot compete in the highly fertile broken ground typical of arable fields or market gardens. Docks are archetypal weeds that grow vigorously without mycorrhizae and are massively invasive in freshly cultivated fields and for some years afterwards, but if the soil is then left undisturbed they tend to fade away - unable to compete with the variety of mycorrhiza-dependent herbs that invade as the soil settles down and becomes less hyper-fertile.

Mycorrhizae have not fully entered agroecological thinking. But their time is coming.

All this adds to the case for minimal cultivation, or indeed for zero cultivation: very little digging and no ploughing at all. We will say more on this in Chapter 6.

A rich organic soil offers one final bonus: it supports a rich assortment of invertebrates - of which, perhaps, the most important are the world's many species of earthworms: more than 6,000 in the world as a whole; 27 in Britain. Some species drag organic material from the surface down into their burrows. Others burrow deep and long, horizontally and vertically, providing precious drainage channels. All of them are great mixers of soil: wonderful cultivators, yet not disruptive like a plough or a spade.

They are food for all manner of birds (including chickens!) and mammals including badgers and foxes, as well as the specialist moles - which horticulturalists tend to abominate, though their tunnels provide excellent drainage.

No matter how fertile the soil, crops will not grow if there is no water, or too much of it. Drought and flood seem bound to grow more extreme and frequent in the decades to come as the global climate changes - so what can farmers do?

## A word on water

At the height of England's floods in the early spring of 2014, Prime Minister David Cameron promised more sandbags. But this didn't and doesn't quite seem to meet the case. All countries, and certainly the more crowded ones like Britain, need to take flood as seriously as the Dutch, forever warding off the North Sea with their dikes, or the Chinese capturing the monsoon rains with their terracing; and to take drought as seriously as did the Moors in Spain, or the Incas, or the people of ancient India, with their underground cisterns and aqueducts and intricate irrigation channels.

Whatever is done (or not done) on the grand scale, individual farmers will continue to suffer from too much rain or too little; and although they can't influence the weather, they can, like the Moors and the Chinese, do a very great deal to control the water that does fall on their land - to take advantage of it, and to reduce the damage it can do. Britain's farmers would surely have suffered far less than they have over the last few years if only they had followed a few simple principles - which, as I saw from my own travels, many did not do.

There are four main approaches:

### *Topography*

The water that falls on the farm in the form of rain is always just passing through. The farmer can hurry it on its way with various forms of drainage, including ditches and the porous clay drains known as mole drains; or else try to slow its escape to reduce the chances of flood elsewhere and as a store for future use. Preferably, the farmer does both at once: seeking both to avoid the damage from too much, and to conserve what should be a precious resource. Both are evident in supreme form in Southeast Asia, where farmers grow abundantly throughout the year on the monsoon rain that largely falls all at once. They achieve this largely by terracing, trapping prodigious quantities of water - enough to raise the semi-aquatic kinds of rice - on hillsides; one of the world's greatest feats of engineering. This has presumably evolved over time not by some individual genius like Archimedes or Brunel but by the people collectively. Farmers worldwide can emulate the principle by ploughing (if they plough at all) along the contours, so that each furrow is a water trap; or by creating swales (barriers laid along the contours). As Vermont farmer and teacher Ben Falk describes in *The Resilient Farm and Homestead,* swales can be made from whatever is around, like straw bales or logs, covered over with soil. Farms these days tend to lack ponds - once a ubiquitous feature, serving many purposes (including aesthetic). If there are enough people on the farm to justify it, it is worth creating a reed-bed. All of these methods are ancient but tried and tested, and very effective. Yet in today's industrial farming they are all but abandoned.

### *The soil*

As we have seen, soils with too much clay tend to dry like concrete in time of drought - then shrug off the rain that does eventually fall, as concrete does; while soils that are too sandy refuse to hold water at all. The antidote to both is more and more organic material, providing sponginess and bringing in the microbes and the earthworms that reduce mineral content to crumbs.

### *Plant the right plants - not forgetting trees*

Crops must be of the kind that are adapted to the conditions - sorghum where it's very dry and hot, grass where there's plenty of rain - but good planting can also help to improve the environment. For example Oxfordshire pastoralist Lyndon Cornwallis has a particular fondness for chicory, a relative of the daisy: its deep roots reach down to the sinking water table in times of drought and provide good grazing when all around is failing. When the roots die, the tunnels they leave provide drainage. Trees, similarly (if they are of the right species in the right places with the right layout) ,bring up moisture from the depths and provide more congenial conditions (and grazing animals are also browsers if given the chance). Trees also reduce flood, holding up the rain at it falls so it does not land in a torrent, and drying the soil after a downpour by transpiration. Here is yet another reason to take agroforestry seriously (see Chapter 10).

### *Micro-irrigation*

Only three per cent of the world's water is fresh and plants won't grow without it - but at present the world is using far more than is replaced by rains, drawing on ancient lakes and aquifers that were laid down many thousands or millions of years ago. Seventy per cent of the water we extract is for irrigation. Some of that is well worthwhile but much of it is pure profligacy: used, for example, to raise soya in vast areas of semi-desert in South America to feed to European cattle who don't need it (see Chapters 7 and 8). Worse: when water for irrigation is run through open channels or stored in open reservoirs much of it, sometimes more than half, may evaporate before it reaches the crops. Though the irrigation systems of many an ancient civilization were indeed wondrous, they too could suffer from this; but the ancients also sent water through tunnels (or bamboo pipes), which largely reduces the problem, and then stored it underground in cisterns. They knew what they were doing.

Here is yet another instance where high tech can come to the rescue. Perforated polythene tubing, carrying just the right amount of water to precisely where it is needed, can be a godsend.

Once we have fertile soil and a steady supply of water then, given a little sunshine, all the many forms of farming are open to us. We will look at the main kinds in the next few chapters.

CHAPTER SIX

# The staples: arable

Arable produces the world's staple crops en masse: not plant by plant as in horticulture, but field by field. Staples by definition provide macronutrients - *most* of our food calories and *most* of our protein. All agriculture is vital but arable can claim to be the most vital of all. We have to get it right. Governments like Britain's, and most of the people with most influence, think we are getting it right already but alas, emphatically, this is not the case.

Of all forms of farming, arable lends itself most readily to industrialization, which creates and concentrates wealth and is therefore equated with progress. Inputs are high - at least of industrial chemistry and engineering; labour is reduced to a minimum and preferably to zero. The whole exercise is practiced on the largest possible scale - as we noted in Chapter 3 there are arable farms in Ukraine of 300,000 hectares and plantations in the Cerrado of Brazil that measure hundreds of square kilometres, where each square kilometre is 100 hectares. It's a fair bet that some Brazilian holdings are as big as *all* of Britain's agriculture (around 20 million hectares). Capital investment is enormous and so, when all goes well, are the profits, at least for those who have shares in it. The accent is on production. Fifty per cent more by 2050 is the target - although as described in Part I the reasons behind this ambition seem almost entirely spurious (and the vast Cerrado estates - raising sugar largely for biofuel, and soya for European cattle - have almost nothing to do with solving the world's food problems).

The results can be impressive. Fertilizers boost yields spectacularly when applied to crops custom-bred to respond to them - so wheat in Britain now averages eight tonnes per hectare, about four times as much as a century ago. Weeds, pests, and all manner of pathogens can be zapped by all manner of agents which could and should become ever more specific in their action and so less likely to damage the surroundings and other creatures. This, at least, is the theory - although in practice this ideal has never been reached and the populations of invertebrates in Britain and surely the world over has plummeted these past few decades, taking many of the birds with them, plus Britain's iconic hedgehogs. The world's favourite pesticides, the neonicotinoids, which attack the nervous systems of insects causing disorientation, paralysis and death, now seem to be killing honeybees and a great deal else besides, in addition to the pests they are intended for. Now, industrial agriculturalists claim that farming as a whole, and particularly arable farming, will be made more efficient by genetic engineering (though they prefer the vague euphemism 'genetic manipulation'). But again the reality falls far short of the hype. The zeal for GM is leading us away from systems that really could serve us well, and the net effect of GM is surely destructive. (For people in positions of influence, Steven Druker's *Altered Genes, Twisted Truth* should be compulsory reading. The tale it tells of corruption and stupidity is chilling - and deadly accurate.)

Nonetheless, despite the perceived drawbacks of industrial arable, the world now produces enough staple food to meet our need for macronutrients twice over and that, compared to the succession of famines described in the Old Testament, and still endured in the mid-twentieth century, is a huge advance. Indeed, the successes seem so striking that many observers, including some who otherwise seem sympathetic to enlightened agriculture, have suggested that we, humanity, should simply acknowledge that industrial, high-tech arable farming really does work, and leave the corporates and their technologists who now control it to get on with the job. The principles of agroecology, they argue, should in effect be confined to livestock and horticulture.

Certainly we should not throw promising babies out with the bathwater. As I have suggested, agroecologists could make good use of at least some of the technologies that have emerged over the past 200 years, not least in arable. Organic farming is the gold standard but absolute adherence to its rules may not *always* be the wisest course. Many farmers worldwide acknowledge that organic methods must be the default position - what should be done as a matter of course unless there is a very good reason to do something else. Yet as we have noted, some of them also argue that *sometimes* a well-targeted herbicide, say, may do less harm than the repeated mechanical weeding that would otherwise be necessary. Similarly, some argue that it could be judicious to add artificial N fertilizer at key moments, just to boost the crop in the spring, say, when rainfall is good and the sun is shining but N is low. Others insist that if the soil is truly thriving with plenty of organic material of the right kind and plenty of mycorrhizae then this should never be necessary. Here is yet another instance where we need more data - though perhaps there is no final answer. In the end, some will argue - legitimately - that 'With best practice industrial chemistry is superfluous - so its only purpose is to disguise sub-optimal practice'. But others might reply, with equal justice, 'Ah, but in the real world best practice is not always possible, and we shouldn't be too purist!' We certainly should *not* say, 'But if I whack on more fertilizer or pesticide I can increase yields with less labour and so make more money. All that matters is "the bottom line"'. But these days that is precisely what farmers are encouraged to believe.

Similar argument surrounds the Green Revolution (the use of genetically modified seeds in developing countries) of the mid-twentieth century.

It was brought about primarily by advanced plant-breeding techniques (though *not* by genetic engineering as has sometimes been claimed!) to produce short-strawed - semi-dwarf - varieties of wheat and rice. Some see this as win-win. After all, the semi-dwarf varieties put less of their energy into making straw and more into grain so this increases the harvest index - the ratio of useful crop to what some farmers (alas) treat as waste. Secondly, traditional varieties of wheat were tall - in Bruegel's rural scenes the wheat is taller than the peasants. If such wheat is heavily fertilized to increase yields it grows very tall indeed and then falls over, or 'lodges', and so is wasted. Semi-dwarf varieties can be heavily fertilized without lodging and so yield even more heavily. Accordingly, the Green Revolution has been hailed as a world-saver. Many argue that the world would have starved without it.

There are downsides, however, as described in particular by the Indian scientist-campaigner Vandana Shiva. Since the semi-dwarfs *can* be treated generously with artificial fertilizers, that is what was and is done, willy-nilly, because more output generally means more profit and profit is the driver. Traditional farms, which typically were mixed and of small or modest size, were and are combined and transformed into monocultural industrial estates which, for example, require far more water and protection from insecticides and fungicides. Then again, small farms are labour-intensive while high-tech industrial estates tend to employ as few people as possible - and though the official rhetoric tells us that this increases efficiency it really means that millions of people are disenfranchised and thrown out of work, to add to the billion who already live in urban slums. Unemployment is the royal road to the poverty that Western governments claim to abhor.

Or then again, the British-based Canadian biologist-turned-arable farmer John Letts points out that the old, tall varieties were very resistant to weeds because they outgrew them, and shaded them out; while the new short kinds are easily overwhelmed by weeds and so need protection using herbicides. John also knows, because he grows them, that the old-fashioned varieties grow best in soil that is *not* highly fertile (high in N) and there are advantages in this; since adding nitrogen is expensive and can lead to pollution. Furthermore, if we were able to grow wheat (and other crops) on soil that is infertile by modern standards, we would hugely increase the area that could be cropped, since most of the world's soil *is* infertile. The increase in area should not be at the expense of wilderness, of course, like the soya on the Cerrado. We could, though, rotate low-fertility arable with existing grassland. So with the old-fashioned varieties we could grow wheat in all kinds of nooks and crannies that are now written off. John himself has shown this by growing what he calls heritage varieties on odd patches in various locations in England and Wales. Because they make good use of mycorrhizae, yields from tall, traditional varieties of wheat in relatively infertile soils are not far below those of modern organic farms, and are eminently sustainable. John points out that wheat straw has a great many uses in its own right - not least for thatching, if it's long enough (he is a thatcher, a miller, and a baker too. Cereals are his thing).

So although industrial farming is in many ways impressive, the drawbacks are manifest too - and we should always ask what *might* have been achieved if we had used our knowledge and skills to enhance traditional practice, and had not simply swept them aside to make way for industrial chemistry and big finance. Most obviously, industrial arable farming is not sustainable: it is squandering finite resources and the collateral damage is enormous, from destruction of soil to climate change to a drastic reduction in genetic diversity and of rural communities. Industrial arable does produce more than enough staple food for all humanity (despite the calls from on high for more and more) but it clearly does not deliver it to everyone who needs it. In its present form, industrial arable agriculture flouts absolutely the vital principle of food sovereignty. Power becomes ever more concentrated in fewer and fewer hands, which is the precise opposite of democracy - the principle for which we still fight wars, or that, at least, is the official excuse.

Surely it would be very foolish indeed - suicidal even - simply to hand over arable farming to the powers that be, the oligarchs, with an invitation to carry on as usual. Certainly it is harder to apply the principles of enlightened agriculture - those of agroecology and food sovereignty - to arable farming than it is to horticulture or to livestock. But we must do it nonetheless. If the agrarian renaissance does not reclaim arable, then it will be forever marginal.

So what does arable farming entail - and what does it look like in organic form?

## The arable crops

By far the most important of the crops that lend themselves to arable farming are the big-seeded grasses known as cereals. Wheat and rice are the biggies. Maize until recently was an honourable third, the basis of serious cuisines worldwide, but now it is mostly hijacked for animal feed and for various industrial purposes including biofuel, which can properly be seen as a scam, albeit a lucrative one. Barley, oats and rye also have an honourable presence more or less the world over, and all deserve to be taken more seriously in the modern diet. Triticale, first bred in the late nineteenth centuy, is an artificial inter-species hybrid of wheat and rye that fills various, mostly subsidiary, roles. Sorghum is the standard cereal of the dry tropics when maize fails, and millet in its various forms is even more heat-resistant. Sorghum makes excellent beer. (Sorghum should not be turned into yet another commodity crop but sorghum beer should surely be exported. There are times when it definitely hits the spot.) Wild rice, a semi-aquatic grass traditionally harvested by Native Americans, and Ethiopian teff, both have local importance. Worldwide, in places where plants can grow at all, there always seem to be appropriate cereals, all of them multi-purpose and deserving a central place at least in the local cuisine. The range of cereals might be extended still further, but the line-up is already wondrous. The American biologist Jared Diamond suggested in *Guns, Germs, and Steel* that civilization could never have got going unless our ancestors had discovered cereals and although this may be somewhat exaggerated - there are other staples too - they have certainly done a great deal to shape the course of civilization. More broadly, the grass family, Poaceae, is one of a shortlist which over the past 100 million years or so has shaped the entire terrestrial ecosystem.

The non-cereal staples - which might have filled the arable niche if cereals had not existed - include relatives of beets such as quinoa and amaranth, and the relative of dock known as buckwheat. All have long been locally important and all are acquiring a global presence. (Buckwheat noodles and pancakes are a great treat.)

Rivalling the Poaceae in ecological significance these past 100 million years or so is the Fabaceae, the nitrogen-fixing legumes. The big-seeded kinds that we eat - the legumes' answer to the cereals - are collectively known as pulses: peas, beans (of various genera), lentils, cow-peas, chick-peas, pigeon-peas. Also of huge significance are the legumes that are not grown primarily for their seeds but for their leaves, which make fine animal feed, and for their ability to fix nitrogen: clovers (*Trifolium*), vetches (*Vicia*), medicks (*Medicago*), alfalfa aka lucerne (also from the genus *Medicago*) and the rest. All in all the Fabaceae complement the Poaceae beautifully, very much to our benefit. Pulses and cereals enhance each other's protein, as outlined in Chapter 2, and the two families together are the key to organic arable rotations, as we will see.

Then there is a range of arable tubers from various botanic families - root tubers like turnips, cassava, yams and various beets, including fodder beet; and stem tubers such as potatoes. Oilseeds are very important, too, of which the main ones in temperate countries are sunflower and rapeseed. Soya is also officially classed as an oilseed, though it is best known as a prime source of plant protein. Groundnuts, alias peanuts, are also of the family Fabaceae. They are wonderfully drought-resistant - more so even than millet - and are very significant oilseeds, but (like soya) are also rich in protein and in some cultures are a virtual staple. (The other principal oilseeds are olives and palm oil but they are plantation crops rather than arable.) Finally, a wide range of horticultural crops are also grown on the field scale, including leaf brassicas and carrots.

The challenge is to raise all these crops, and especially the staples, by the methods of agroecology. They can all be raised on the small scale, as horticulture, and this is well worth considering, as discussed in Chapter 9. But how can we raise them on the field scale without compromising the principles of agroecology?

Well, even though agroecology does not always mean strictly organic farming, organic methods are certainly the basis. So we need field-scale organic.

## Organic farming begins with the soil

All good farming begins with good soil as discussed in Chapter 5: structure, fertility, moisture - not too little and not too much; a comfortable balance of water and air. Here I want to look at the specific contribution of the organic content - the bit that the conventional, industrial farmers of today have so often, shamefully, neglected.

Just to recap: organic material contributes to soil structure - to a large extent it *is* the structure - by supplying carbon, C, in many and various complex forms, but mainly in the form of cellulose, the chief stuff of plant cell walls, in various degrees of degradation. Organic material can also contribute a whole range of nutrients but the one that is needed in greatest amounts, the essential ingredient of proteins and nucleic acids (and more besides), is nitrogen, N.

In some forms - peat, bark, leaf mould - organic material contributes a great deal of C, but very little N. So any of these may greatly improve the soil's structure and its ability to hold water and provide the sponginess that can let the air in, and so enhance the growth and health of the plants. But they will not, in these forms, provide nutrient, and in that sense make the soil more fertile. In other forms - slurry, chicken manure, decaying animal flesh (such as fishmeal or blood) - organic material may provide a great deal of N but little C. Then it nourishes the plants in the immediate term but contributes little or nothing to the structure of the soil.

Very few plants apart from the highly specialist sphagnum mosses do well on pure peat (and they presumably get their N and other nutrients from incoming insects and debris). In practice, on most if not all of the soils that are used for agriculture, it is more or less impossible to introduce too much C. But it is certainly possible to introduce too much N - although there has to be enough for optimal growth; and what is optimal depends on the species of crop, the stage of growth and the weather (if it's too cold or too dry then temperature or water become the limiting factors and plants can't grow however much N they may be given).

So the task for organic arable farmers and growers is to maintain soil C and preferably to raise the level year by year to the maximum the soil can contain (though carbon does fluctuate naturally as the seasons pass); but to regulate soil N so that the amount at any one time is adjusted to the needs of the crop - not too much so the surplus is wasted, or too little so the crop is held back. Need varies with the seasons, too: so wheat appreciates a boost in N fairly late in its growing cycle when it is focusing on swelling the grains. In truth, though, a late-season burst of N serves mainly to raise the nitrogen content of the grain - which with wheat tends to mean the gluten content. Modern commercial millers and bakers require this, because it's the gluten than enables the bread to rise. But more traditional forms of bread, such as sourdough, do not require such high-N wheat; and cereals (notably barley) intended for brewing should in general be low in nitrogen.

Whatever the complications, the question is: how can organic farmers ensure that the cereal crop has all the N it needs (though not too much) throughout the season? They cannot simply dump a load of manure on top of the crop without destroying it. The obvious time to apply manure or compost is in the autumn when the main crop has been harvested and the field is temporarily bare - but these days it is not considered good practice to leave bare fields in winter, especially with manure on top, because the precious N leaches away with the winter rains, and an asset becomes a pollutant. The aim as far as possible is to ensure permanent plant cover.

None of these problems can be solved absolutely (there are built-in contradictions) and they certainly can't be solved by any kind of universal formula. But there are some useful and well-established principles. Chief of these is rotations.

## The absolute importance of rotations

Rotations vary enormously from farm to farm and from decade to decade - but all have key features in common. Notably, periods in which fertility is raised (which mainly means adding N) alternate with periods in which crops are grown in the fertile soil. In the productive period, the fertility steadily falls, so the crops that benefit most from high N are grown first, followed by those that need less, and so on.

The fertility-building phase is key: as Stephen Briggs comments in *Organic Cereal and Pulse Production*, it is the cornerstone of organic arable farming. In all-arable organic farms the fertility must be built entirely by N-fixation, courtesy of legumes. Much-favoured in Britain are red clover, which is tallish and upright and quick to establish; or the sprawling white clover, slower to establish but more enduring. Both are of the genus *Trifolium*. Then there are various vetches, *Vicia* - from the same genus as the broad bean, which in various forms manifests as the field, tic, or horse-bean; lucerne (alfalfa), genus *Medicago*; and sainfoin, genus *Onobrychis*. All of the above are grown for their leaves. But year by year, as N increases in the soil, the rate of fixation falls off - for, says Briggs, 'Legumes fix significant amounts of N only if they cannot get it from the soil.' But if the farmer takes a crop of N-rich grass and clover for silage, the soil N goes down and the remaining legumes may respond by fixing more N - so that's a bonus. N removed in the silage can be added again after it has passed through animals and re-emerged as manure. However, if slurry or poultry manure - with very high N relative to C - are added to the ley, fixation goes down. But well-rotted cow manure adds P and K (and carbon) but contains less N than slurry, so it can bring net benefits.

Commonly, clovers or other legumes are sown at the same time as the cereal seeds - that is they are 'undersown'; and so they supply some extra N to the newly emerging cereals. This is especially appropriate when winter wheat is sown in the autumn, leaving time for the clover to grow before winter sets in, and ready to provide a boost in the spring when the wheat crop should be growing most rapidly. Some farmers sow wheat or other cereals directly into fields of established clover.

A range of other crops from various botanical families may also be slotted in to the rotation. In Britain these include white mustard, stubble turnips, kale, fodder radish and forage rape from the brassica family, Brassicaceae; phacelia, a relative of borage and comfrey in the family Boraginaceae; perennial ryegrass and forage rye, of the Poaceae; sunflowers from the daisy family, Asteraceae (formerly known as the Compositae); and buckwheat, of the dock family, Polygonaceae. These 'break' crops (secondary crops grown to interrupt repeated sowings) serve several purposes. Some of them can be pressed into service as animal feed. Sunflower and buckwheat are potentially saleable. Most of them are different botanically from the main arable crops and so interrupt the life cycles of the principal pests and diseases. Some bring extra benefits: a cover of white mustard helps reduce scab in following potato crops. All, by covering the ground, help protect the soil from wind and rain. But they also continue to transpire through the winter and so help to dry the soil - which in turn allows it to warm up more quickly in the spring. Perhaps most importantly, all may be treated as green manure - ploughed in or left on the surface as mulch to improve the soil. Because such crops do not fix nitrogen, they do not add to the soil N - but they do bring it up from the depths and act as a store. So long as the N is contained within a plant it cannot be leached or sublimate away, and when the plant material is re-incorporated, all is restored. Break crops also photosynthesize and so fix carbon, and it is always good to add organic carbon.

With fertility restored, the next few years of the rotation can be used to raise crops - to be consumed locally, sold for money, or fed to on-farm livestock; although the farmer can continue to boost fertility during the cropping phase, to some extent, by undersowing with legumes.

Variations on the theme of rotation are endless - they depend on the soil, the topography, the climate, how the fields have been treated in the past, the local history of weeds and diseases, the state of the market and the needs of the farm - and the preferences of the farmer. Crucial, too, is whether or not the farm has livestock. But whatever the variations, there are a few key requirements. First, the farmer needs to strike the right balance between building fertility and raising crops. If fertility is not built up enough, then the crops suffer. But if too many years are spent building fertility then the extra time is wasted and growing time is reduced for no good reason. The fertility-building phase should be followed by a crop that can respond most usefully to the added fertility, followed by crops that are successively less responsive. Successive or undersown crops should be botanically different so they don't spread their pests and diseases one to another. An appropriate time gap should be left between any two crops of the same species to minimize the spread of disease from one to the next ('carry-over'): two years between wheat crops; three to four years between oats; five to six years between crops of red clover; and so on. The ground rules are commonsense biology but they all add to the complexity. Finally, it can be helpful on all but the smallest units to divide the arable farm into blocks, with each block following different rotations or - more likely - all following the same rotation but out of phase with each other, like musical canon. That way the farm maintains a steady output of each kind of crop, which is good for cash flow and for customers, and it is also easier to spread the labour.

Since there are so many variations none can perhaps be said to be typical; but here is a five-year version favoured by John Letts which he devised for his 'heritage' wheats. It begins with a year of peas and beans - mixed - which produces excellent animal feed and adds N to the soil. This is followed by two years of genetically diverse heritage wheat - mixtures of traditional, including ancient, varieties. The first wheat crop, says Letts, might perhaps be winter-sown, and the second spring-sown, and both should be undersown with clover. In both years, the straw left after the grain is taken is left to decay, and worked into the soil. The two years of wheat would be followed by a crop of rye, which is good for human consumption, or by triticale which isn't, again undersown with clover. In the last year he would plant oats mixed with peas to produce another fine animal feed - of a kind known in medieval times as bulimong; or, perhaps, barley. Then it's back to the beginning.

If the farm is divided into blocks then that becomes the sequence for the first block; the second runs the same cycle, but starting with triticale; the third one starts with beans; and so on. Five years, five blocks. Of course, though, if a particular block does not begin with clover then we have to assume that there must be some residual fertility to enable the cycles to begin at all. But whatever the sequence, the organic arable farm as a whole should never have less than 20 per cent legumes, and if the farm is not divided into blocks it may be 100 per cent legume. There may be no cereals at all in any one year and there should never be more than 60 per cent. Contrast this with industrial arable farms which aim for 100 per cent cereal every year. But nothing in farming is simple, for as John Letts says: 'I think it is possible to have continuous cereal cultivation on much of an arable holding every year, as long as you interplant and undersow with legumes. But the key is to have a cereal crop that is suitable for these situations.'

When the farm has livestock we might envisage a six-year rotation that begins with a grass clover ley that is cut for silage, with the aftermath grazed perhaps by chickens and pigs. Year two is much the same. In the late autumn of year two, wheat is sown, to stand through the winter and occupy most of year three; though followed at the end of year three with a crop of *Phacelia* (the relative of comfrey). Year four is devoted to peas or beans, with triticale sown at the end of it to stand the winter and occupy most of year five. This is followed later in the year by a ground cover of vetch or mustard which grows or at least survives through the winter - and barley is sown through the existing ground cover in the spring of year six.

It's worth noting that the earliest farmers did not rotate their crops at all. They planted them in the same fields year after year. But then, the early types of wheat were adapted to this. Fertility and therefore yields were maintained, we can now see, not by rotations but by mycorrhizae. Industrial farms in general do not rotate because they contrive to make up for all nutritional deficiencies and the build-up of pests and diseases with industrial chemistry.

That then, in barest outline, is the method of agroecology: what organic arable farmers aspire to do, and why. So, what of the future?

## Towards the future

The neoliberal industrialists who run the world are convinced that we need more of what we have now: more and more output (even if the output is designed to be fermented and then burned as 'biofuel'); more and more capital investment (meaning interest, meaning circulation of wealth); bigger and bigger farms (economies of scale and more and more power to the people who call the shots); and husbandry simplified to suit the needs of heavy engineering and industrial chemistry which is where the biggest profits are, at least on the production side - which means monoculture, often culminating in cloning which implies absolute genetic uniformity. GM is applied wherever it can find a foothold, not necessarily because it adds anything truly useful, but to ensure that a few high-tech companies and their supporting governments have more and more control. Of course the collateral damage is enormous but that, we are told, can be cured by more high tech which is also lucrative (and also serves to concentrate power).

If we want farming that is actually designed to provide good food for everyone without wrecking societies and the world at large then, as is so often the case, we need to do the precise opposite of what those with most power consider to be progress, which they use our money to promote. Some farmers and some institutions are already on the case.

## The quest for diversity

We need to reverse the drift or rather the drive towards uniformity: fewer and fewer varieties within each crop, and greater and greater genetic uniformity within each one. Varieties of crops and breeds of livestock do need to be consistent: that is, they have to have a recognizable set of characteristics to guarantee (as far as possible) that they will do what the farmer and the consumer wants and expects; and crops should, ideally, breed true - produce offspring that are the same variety as themselves - so that the seed can be saved. But many crops - like runner beans, for example, cannot be genetically uniform because they are out-breeders: they need the parents to be genetically different. The grower must put up with some variation within each variety, though this is minimized by controlled hybridization between fairly uniform (meaning more inbred) parents. Wheat, however, is an in-breeder - self-pollinating, and more uniformity is possible without hybridizing.

Because uniformity is potentially dangerous (it opens the door to disease), there are moves afoot to increase the within-crop variation deliberately. One is practised by John Letts: he simply plants mixtures of traditional varieties together in the same field. The resultant grains are very mixed and the flour that results can produce some excellent bread and cakes, though the baker has to adjust his or her technique to match. Martin Wolfe at Wakelyns, Suffolk (see Chapter 10), has a different approach: he plants judicious mixtures of phenotypically (physically) similar but genetically various modern plant varieties together, as a mixture. The idea behind both approaches is that in a dry year, say, the types that are more drought-tolerant will thrive at the expense of the ones that like more water; and in wet years the moisture-lovers will predominate. Although in any one year the mixed crops may yield less than a monoculture of a variety that happens to suit the weather, over a decade or so the *average* yield from the mixture should be higher. Martin has shown at Wakelyns that this is indeed the case - and since the grain is grown organically, chemical inputs are zero. Finally - which may be considered a bonus: although wheat is primarily an in-breeder it is not exclusively so and over time there should be enough cross-breeding within the mixed crops to produce new landraces: populations that are selected in the field to cope with the particular local conditions, but with enough in-built variation to enable them to cope with fluctuating climate.

## Cereals in horticulture

Cereals are favoured in the east of Britain partly because it's drier, sheltered from the Atlantic-borne rains by the Pennines and the Cairngorms and the rest, and partly because, as Noel Coward remarked albeit with some hyperbole, East Anglia in particular is terribly, terribly flat. So East Anglia has become more emphatically arable as farming has become more industrialized, favouring bigger and bigger machines. The scale in East Anglia falls far short of Ukraine but 2,000-hectare farms with fields of 100 acres plus are already commonplace. The buzz has got round in arable even more than in dairying that only the vastest scale will do.

But it just isn't so. Cereals can be grown on the small and indeed on the horticultural scale. It would be good if, say, two-hectare market gardens or smallholdings also had two hectares of cereal, and the two could then alternate, each becoming part of the other's rotation. This would double the reason for adding poultry and/or pigs and becoming bona fide smallholdings. There would be little point in growing standard cereal varieties in standard ways but plenty of point in growing rarer varieties, perhaps in mixtures. Add a small bakery or brewery (or, better, several such units feeding into a small bakery or brewery) and we have a most interesting set-up with plenty of scope for varied employment; an ideal small-to-medium-sized enterprise for a community to own and run. Furthermore, once small plots are again brought back into the arable fold, the cause of self-reliance is greatly enhanced, in the world at large as well as in Britain.

The precedents are out there, albeit often lost, or nearly lost, in the mists of history - though still to be seen in essence in enterprises like John Letts' or in small farms, not least in Eastern Europe. It's just a question of doing it.

## Perennial arable?

Arable farming poses some severe dilemmas. Because the crops are annual (at least in effect) they must be resown every year; and arable crops are immensely susceptible to weeds - plants which, by definition, like disturbed ground which they prefer to be rich in nitrogen. Traditionally, resowing means cultivation which on the grand scale means ploughing. Even on the small scale, with horses instead of tractors, this disturbs the soil biota (soil organisms) and exposes soil carbon to oxidation, while bare soil can erode through wind and water. On the grander scale, with heavy machinery, the soil may be compacted, which on clay soils in particular causes pans - highly compressed layers which neither roots nor water can penetrate, so effective soil depth is reduced, deep nutrients are lost, and wet spots build up. Methods to reduce the need to plough by minimum tillage or 'min-till' tend in practice to rely on herbicides to kill the weeds which is also undesirable. Organic farmers seek to cope with weeds by timely rotations but organic min-till is logistically difficult.

One possibility, now being worked on, is to produce perennial cereals that don't need to be replanted every year. The traditional approach is to cross existing domestic cereals with compatible species of wild grasses that are already perennial, and then breed back to produce a plant that combines year-by-year persistence with big seeds. Another is to identify the genes that confer perennial status and transfer them directly by genetic engineering. Both approaches are being tried. The GM route may sound attractive but precedents with other GM crops and a great deal of biological theory suggest that the problems will be far greater than can now be fully anticipated. In general, regarding perennial cereals, we can only reasonably say, watch this space.

## Silvo-arable

One approach that is already vindicated is the silvo-arable route: growing cereals between rows of trees. Martin Wolfe at Wakelyns has shown that it can work very well indeed. But for traditional (industrial) arablists, trees have been an anathema. More in Chapter 10.

## New baking

Finally - a key theme throughout this book! - enlightened agriculture of all kinds, including organic, mixed-cereal horticulture, cannot work properly without a corresponding food culture (as outlined in Chapter 11). For new-style enlightened arable to flourish, we need bakers and brewers who are prepared to experiment with new ways of cooking and making beer, all on the small scale; and communities to support them - provide the start-up finance and the customer base; and people at large (the word consumer is hateful but there doesn't seem to be another one) who care enough about food and all it stands for to insist that food is produced in the right ways and who will buy it when it is grown in the right ways. We really do need to relearn how to cook. Good cooks lead good farming.

In short, arable in the end is the jewel in the agricultural crown; what most people mean by farming. We really cannot surrender it to the neo-liberal industrialists. We really must keep it within the enlightened fold. The possibilities are all out there. In arable, as in all farming, these are exciting times.

CHAPTER SEVEN

# Livestock I: The basics

Through all the ages, all kinds of people for all kinds of reasons - ecological, nutritional, moral and metaphysical - have argued that we should not raise animals for food. It can be a damn' near-run thing, yet on balance I am sure that livestock farming can be justified.

Certainly a hectare of cows on fertile land produces less food energy and protein than a well-grown hectare of wheat - but properly organized mixed farms with animals and crops are the most productive of all; and *most* of the world's farmland is not suitable for arable; if we didn't use it for grazing we wouldn't be using it at all. All in all, well-balanced mixed farming takes up less space per unit of food produced than dedicated arable or horticulture, and can accommodate more wild birds and so on. Modern-day industrial livestock farming is certainly profligate - but mainly because it is geared to money, and while oil is still plentiful it pays big business to feed more than a third of our staple foods to animals, and then burn some more and call it biofuel. But if we really cared about the biosphere we could produce all the meat that's needed for sound nutrition and great cooking (see Chapter 2) with just a little cereal or pulse to supplement more natural feed. Too much meat of the kind that's been stuffed on concentrate evidently can lead to heart disease and a great deal more, but smaller amounts of animal flesh that has been raised at a sensible pace on a natural diet may actually benefit the heart and arteries. Animal husbandry (management and care of farm animals), can be cruel, but it doesn't have to be.

So livestock farming passes the principal tests - provided we do it well, with maximum concern for the welfare of the animals and the wellbeing of the biosphere, and with empathy for both; and this is best achieved, as is always the case, through a judicious balance of traditional practice, evolved and honed by many millions of farmers over many centuries, and the insights of modern science. As always, the problems with livestock farming really begin when we leave the control of it to people who are focused primarily or exclusively on the creation of short-term wealth, and on their own careers.

So what does livestock farming look like when we take serious matters seriously?

## The basics

'Livestock' includes fish and molluscs and crustaceans (and these days even insects) but here I'm concerned only with warm-blooded land vertebrates. Zoologists divide them into birds and mammals, but for farming purposes they are best divided into omnivores and herbivores.

### *Omnivores*

'Omni' means 'all' and 'vore' means 'devour' so omnivores eat both flesh and plants; and the most important ones by far in agriculture are poultry and pigs (and dogs are significant in some countries). Pigs and poultry would be happy to eat what we eat, though both clearly derive some energy and protein from grass. For some species of geese and ducks, grass are a staple: domestic geese were traditionally raised from May until Christmas entirely on pasture (as the newly founded community-owned farm in my own Oxfordshire village is now doing). Some pig breeds such as the Berkshire are known for their fondness for grazing. Some say, though, that the pig's ability to utilize grass is more a matter of upbringing than of breed. Their guts adapt remarkably to what's available.

But even when they do eat what we eat, pigs and poultry seem to have lower gastronomic standards than we have developed and so can derive some or even most of their energy from leftovers, known as swill. That is why traditional cottagers kept pigs in the first place - as waste disposal officers; to provide manure for the vegetables and fruit; and to cultivate the ground (they are great diggers, probers and clearers of weeds). The meat, salted to last through the winter, was a bonus (though a very considerable one). Use of swill is now restricted, which takes away one of the main, traditional reasons for keeping pigs and poultry in the first place. Ostensibly the new restrictions are to control disease but there is a clear commercial interest, since they ensure that pigs must now be fed mainly on custom-grown grain and soya. Thus pigs (and cattle too, alas) become a branch of big-time arable - an excuse for growing more grain. Pigs and poultry can and should be fed cereal, but traditionally they were given only what was surplus.

Kept according to traditional strategy, pigs and poultry are key players in mixed farms and the natural stepping stone between market garden and bona fide smallholding. Today's factory farms, housing up to a million pigs at a time, are ecologically and socially grotesque. They are perceived to be modern but the thinking behind them - big is good, and bigger is better, and we have the technology to do anything we choose - belongs, at best, to the nineteenth century.

### *The full-time herbivores*

Specialist herbivores are able to do what non-specialists cannot do - derive *most* of their energy from the cellulose and other complex carbohydrates that are the stuff of plant cell walls. Cellulose is the principal component of straw and of wood (when the cellulose is toughened with lignin) and as such it is the favourite food of termites, a host of beetles and bracket fungi. But for mammals and birds the main source of cellulose is, and are, leaves and stems - grass in all its forms (including straw) and browse (the leaves of bushes and trees); though herbivores (as well as omnivores) on farms are also fed roots of many kinds, including turnips and a large kind of beet known as mangels. On land, cellulose has the greatest collective biomass by far of all organic molecules, which means that for animals that can cope with it, cellulose is the greatest of all potential sources of food energy on earth (or at least on land). Humans cannot digest cellulose, but omnivores like us and out-and-out carnivores, of which cats are the prime example, gain vicarious access to the vast reserve of cellulose by eating specialist herbivores. Thus the flesh of herbivores is our access route to nature's biggest terrestrial nutritional bonanza.

Yet our domestic herbivores do not digest the cellulose themselves. They lack the necessary 'cellulase' enzymes. Instead, herbivorous mammals and birds rely on microbes - bacteria, archaeans and protozoans - which they harbour in extensions and expansions of their gut. Many of those microbes do produce cellulases, and break down the cellulose and accompanying materials to produce volatile fatty acids (VFAs) which the host animal absorbs and which, when they eat a natural diet, become their chief source of energy. Thus the specialist herbivore is a walking fermenting chamber with an onboard microbial ecosystem of unimaginable complexity. Modern nutritionists stress that our own gut metabolism, and hence our nutrition depends far more on our gut microbes than was ever supposed - though none of our own microbes produces cellulases. If they did, and we could live on grass, history would have run a very different course.

The terrestrial herbivores, whether birds or mammals, are of two main types: the foregut digesters and the hindgut digesters. In the foregut digesters the fermenting chambers are specialist extensions of the stomach. The only foregut digesters among birds (or the only one that I know about) are the peculiar hoatzins of South America, which smell like cows. Among mammals, kangaroos and their relatives are foregut digesters of cellulose (and have at times been semi-domesticated for meat), but the best known and the most ecologically important foregut digesters by far, are the ruminants - a huge group that includes cattle, sheep, goats, antelopes, pronghorns, deer and giraffes. Camels and their relatives are similar and are sometimes called 'pseudo-ruminants'.

The stomachs of true ruminants are divided into four distinct chambers, each with its own specialist function, but the biggest chamber of all - the size of a space hopper in a mature bovine - is the rumen, where the serious digestion takes place. But cellulose is difficult stuff to break down even with assistant microbes, so to do the job properly ruminants have two goes at it. First time round they gobble down the leaves (no time to waste) and pass it to the first two chambers of the stomach (the rumen and reticulum) where it mixes with their own enzymes; then they cough it up again a bolus at a time and chew it thoroughly in their own time - which is known as chewing the cud. Then they swallow the well-chewed pabulum a second time. On its second trip the food passes to the rumen for serious microbial attention, then on to the omasum for further refinement, and so to the abomasum which is equivalent to the human stomach. Hindgut digesters include rabbits, horses, elephants, koalas and most herbivorous birds such as the grouse. The microbial fermenting chambers are formed from an expanded and extended colon and/or from a caecum - a sideways extension of the gut positioned between the stomach and the small intestine.

Vegetation is highly variable and different species and breeds of herbivores approach it in various ways. Some breeds of cattle do very well on low-grade vegetation, although they grow slowly - notably the humped zebu cattle of the tropics and subtropics which in traditional systems typically function on crop residues including cow-pea stalks and wayside rubbish. Indians do not traditionally eat their cattle but - since the husbandry is more or less free - zebus and their various hybrid derivatives pay their way many times over by pulling carts and ploughs, while in Africa cattle are commonly perceived as currency (which is much more reliable than money and doesn't have to be carried around. It walks). Scotland's Highland cattle and other upland breeds, raised primarily for meat, also do well on rough pasture - upland fog grasses and heather.

High-yielding industrial dairy cattle by contrast, like the black-and-white Holsteins that now are Europe's leading breed, need rich diets to maintain their milk output - lush high-protein custom-bred ryegrass heavily supplemented with concentrate in the form of cereal and soya (and other materials including animal protein). High-yielding dairy cattle also require enormous quantities of water. Thus the import of Holsteins into tropical countries in recent decades in the name of progress has often proved disastrous, as Western intervention generally does when it's in its one-dimensional mode (as it usually is). The native cattle that make good use of what's available are swept aside while the flashy incomers languish and die, so vast sums are wasted while the local people are worse off than before. The interventions from outside that really do bring benefits are often simple and low tech, yet are nonetheless scientific. So urine added to rice straw can produce fodder that is adequate, not only in energy (from the cellulose in the straw), but also in protein (which the ruminants' gut microbes synthesize from VFAs and urea). That is simple but it's good science, with sound physiology behind it.

Green vegetation on the whole is not rich in energy, relative to its bulk, and it can be hard even for a specialist herbivore to get enough energy and protein from a diet that consists mainly or entirely of leaves, especially rough leaves. Warm-blooded animals - like mammals and birds - use most of their food energy keeping warm, and since little animals like mice and finches cool down more quickly than big ones, they must maintain a far higher metabolic rate. So although big animals eat more than small ones (of course) they eat far less *per unit body weight* than small ones do. If small animals eat food with a low energy content they simply cannot eat enough to stay alive. Since leaves have a low energy content relative to bulk, no bird smaller than a grouse and no mammal smaller than a rabbit can live on them. Small plant eaters, like goldfinches, eat seeds, which are far more nourishing. But it is not easy even for big animals to get enough energy from an all-leaf diet. So an elephant in the wild must typically browse or graze for 17 hours a day, while lions with their rich meat diets typically doze for 20 hours a day. Cows and horses with their big bodies and enormous guts are adapted to a bulky, leafy or stalky diet - but if we want horses to work, or a cow to produce six times as much milk as she would in the wild, then they must be given concentrates to supplement the basic diet of grass.

### Milk

For mammals, lactation is the greatest physiological strain of all. A modern commercial dairy cow producing 1,200 gallons (6,000 litres) of milk per year (which is the present British average) must more than double its resting metabolic rate in order to cope; and elite cows these days produce 10,000 litres-plus. In human terms, the energy required by the elites is the metabolic equivalent of the Tour de France - but lactation lasts 10 months while the Tour de France is over in 23 days (including two rest days). Furthermore, gestation in a cow takes nine months and cows are expected to give birth every year, so most of the time the cow is both pregnant *and* lactating. Wild cattle produce less than 300 gallons (1,500 litres) per lactation and may not give birth every year, and so can manage on grass (and browse) alone - as domestic cattle traditionally do. But today's high-tech dairy cattle need a far richer diet: specially bred soft and protein-rich ryegrass; and significant intakes of concentrate, which they are generally fed while they are being milked (among other things it lures them to the milking parlour). Cereal is, and has always been, the main concentrate. As Graham Harvey records in his excellent *The Carbon Fields*, dairy cows these days may get through three tonnes of cereal a year - and only 10 per cent of the energy that goes into their milk comes from grass, their natural food. More than 90 per cent of the world's vast and growing soya crop is fed to animals - largely to industrial dairy cattle. For this, the marvellous ecosystem of Brazil's Cerrado is being trashed, although it could, if conservatively managed, support large numbers of people *and* its native wildlife. The industrialists say soya is essential and many young farmers, born into the industrial age, apparently believe that this is so. Yet cattle have been domesticated for 10,000 years and 99.999 or so per cent of them never had a sniff of soya until the last few decades. In truth, as Harvey says, it's simply easier to dispense pellets from a bag than to manage grass properly. If we do manage grass and browse well, then cattle and other herbivores become a huge asset, for us and the for biosphere. If we just feed them on cereal and soya, and do not manage the grazing, they become a serious burden on humanity and on the world.

Over-rich diets are cruel, too. Cattle are not evolved to digest cereal. Gut microbes plied with cereal produce various extraneous metabolites that make the animals sick. This and other metabolic knock-on effects help to shorten the life of very high-output cattle - adding to the mastitis and the mechanical strains imposed by the vast udder. So Britain's high-flying industrial cattle typically manage only three lactations (at most), while wild and traditionally reared cows commonly manage ten or more. This raises huge issues both of economics and of morality. Should we adapt the diet to the natural physiology of the animal and be content with lower output? Or should we feed our animals with as much rich food as they can stand, to maximize their output? Since ultra-high input with ultra-high output can cause a great deal of suffering, common humanity suggests the former. But when rich food is cheap (mainly because it needs oil to produce it in large amounts, and the price of oil is adjusted to make this possible), accountancy suggests we should push the animals (and the farmers) to their limits, and then some. As things are, accountancy trumps morality. This is true in all fields but particularly, it seems, in modern farming. But there's a twist - for in Britain at least, dairy farmers who have gone down the industrial route, as they have been urged and virtually forced to do these past few decades, are going out of business by the day. Since ultra-productive cows are short-lived they need constant replacement, and replacements are expensive. To achieve economies of scale, farmers take on more and more cows per worker - which means more and more fancy gear, milking parlours and processing plants and all the rest, which means capital outlay which means debt, often of a million pounds or more. So industrialized dairy farmers must earn a fair whack - enough to pay a couple of workers - just to pay the interest on their loans. Indeed, interest on loans, money for the banks, accounts for more and more of the total British food bill, which perhaps is not what most politicians intended but certainly suits the modern financiers. Industrialization leads to centralization. In the UK, six processors (Arla/Express, Dairy Crest, Müller Wiseman, Glanbia, Associated Co-operative Creameries, and Nestlé ) now control 93 per cent of dairy processing. Farmers are given a choice: gear up and plug in, or get out. It's Hobson's choice.

But there is a third way. Harvey persuasively argues that we need, perhaps above all, to return to small dairies. In much of Europe, as in the Austrian Tyrol, small dairies are still the norm, properly supported by government. Britain is hostile to small farmers but even here, some brave souls are showing that micro-dairies can work, including Nick Snelgar in Hampshire, described in Chapter 8. Above all, small enterprises need support from the local community. Small farms can serve big cities too as they did as a matter of course until about half a century ago, and we should strive to revivify suburban and peri-urban farming of all kinds. This is enlightened agriculture in practice: farming geared to the realities of animal biology and the needs of the biosphere, and to the health and autonomy of humanity at large. More on small dairies in the next chapter.

### *Dairy and beef*

The dairy and the beef industries are often conceived as separate enterprises with specialist breeds for each; but some breeds are dual-purpose and in practice *most* of the calves that are turned into beef (and veal) are produced by dairy cows (such as Holsteins, Ayrshires or Jerseys) that have been crossed with beefy bulls (such as Charolais or Aberdeen Angus). In dairy animals, emphasis is on the females; and as the whole enterprise has become more specialized and high tech and farm animals have been reduced more and more to commodities, it has focused specifically on the udder. Today's elite dairy cows are monsters: skin and bone with an udder the size of a dustbin (they are commonly crippled from straddling it, and suffer from both mastitis and ketosis, the result of a too-rich diet). In beef cattle the focus is on muscle. In wild cattle of all species the bulk is at the front - big shoulders and sometimes a hump, as is most conspicuous in bison; very good for seeing off other bulls and building harems. But the best steak, it is widely agreed, is in the hindquarters, so by selecting and crossing, breeders and farmers have produced animals with the weight towards the rear.

Today's ideal beef breeds are oblong. In Britain cattle intended for beef are usually castrated and are then called bullocks or steers, although the Germans tend to favour bull beef. Intact animals grow bigger and faster but they are harder to handle. Females of all but the skinniest breeds make good beef too, but males are bigger and grow faster. Some herds are kept purely for beef. In Britain an important minority of beef calves come from dedicated 'suckler' cows whose only job is to produce one calf per year which each cow raises on her own milk, and the offspring when weaned are commonly sold on to be raised and finished (fattened) elsewhere. Suckler herds are especially favoured in hilly areas where the grass grows well but it is hard to round up dairy animals for milking, and there are few local people to sell it to.

Most beef animals in Britain come from the dairy herd. Cows will not produce milk if they don't give birth first, unless they are plied artificially with hormones (which in Europe, mercifully, is illegal). The calves are taken from the mother as soon as they have drunk their colostrum (the milk that's produced in the hours after birth, full of antibodies) and raised artificially.

The life of the calf then follows one of several courses. The farmer keeps some of the females as herd replacements: the shorter the life of his or her cows, the more replacements are required. To produce calves that will themselves become dairy cows, the farmer crosses the cows with semen from a dairy bull to produce a pure-bred dairy calf. But the male calves and the females surplus to requirements can be raised for veal - not the kind that are raised cruelly without roughage and sold very young in an anaemic state; but raised for six months and then sold as 'rose veal'. Or the males and young females can be raised and fattened for beef, typically at 18 to 30 months old. For beef calves, the farmer generally needs to cross the cows with a beef breed to produce a hybrid. Hereford-Friesian hybrids have been a great favourite but crosses between Holsteins (smaller than Friesians but milkier) and big continental breeds are now perhaps more commonplace.

So the ideal, at least, is to produce female-only calves from dairy-dairy crosses to become replacements; and male-only calves from beef-dairy crosses to produce beef. But in a state of nature, half the calves will be female and half male so about half the calves will be less than ideal - pure-bred dairy males, or female beef-dairy hybrids. Nowadays it's possible to separate the female sperms (containing an X chromosome) from the male sperms (with a Y chromosome) before inseminating, and so pre-determine the sex of the offspring. Here is high tech in the service of sensible farming - which should be what high tech is for!

Sheep too worldwide are multipurpose - kept for meat, milk and wool, and not necessarily in that order. Milch sheep are key animals in much of the world - their milk the source of some of the world's outstanding cheeses, including halloumi (sheep and goats' milk mixed) and the unsurpassable Roquefort. Sheep and goats make it possible to be a dairy farmer even on the smallholder scale. Entire economies have been built on wool, including much of medieval and Renaissance England (which among other things financed some our finest parish churches). Some breeds in Britain are still kept largely for their fleeces (not least the multi-coloured Jacobs and Herdwicks) but the main motivation by far these days is meat. In past times mutton was favoured, from mature animals; and many people still prefer its often powerful gaminess. But most sheep in Britain these days are slaughtered for lamb, which by definition must be less than 12 months old (though the point is sometimes stretched), and most are slaughtered at 10 weeks to 6 months. Most lambs spend their entire lives on pasture.

## What does it mean to be efficient?

Farmers must always try to be efficient - but efficiency can be measured according to various criteria, and it's important to pick the ones that are most appropriate. In the present economy, money rules, so farmers must increase efficiency by maximizing the cash value of their output while cutting costs, in whatever ways the law allows (and it allows a great deal that is undesirable).

Far more important, and always pertinent in any kind of economy, is the concept of *biological* efficiency. This is trickier to pin down but can be measured, for example, by the ratio of food energy or protein put in (in the form of grass or cereal or whatever) versus the amount produced in usable form (as meat or milk or whatever). When farmers try too hard to reduce this ratio the results are cruel - the animals may be fed a too-rich diet and induced to be far more productive than they are physiologically or anatomically equipped to be, and suffer many forms of infirmity. (But their misery need not affect their performance too much, so who cares?)

All farmers, however enlightened, must always keep at least one eye on the money. They must have something to sell and it must not cost more to produce their goods than they get back. The overall task, then, is to increase biological and cash efficiency while working within the animal's comfort zone. It can be a narrow path, but within a halfway sensible economy and with sensitivity, it is eminently achievable.

In practice, whatever the economy, the farmer must be guided by three fundamental principles of biology.

First - without overstressing the animal, and with many a conditional clause - the farmer should strive to minimize the time from birth to slaughter. Partly this is a matter of turnover: the quicker the throughput, the greater the output in a given time and space. But partly this is a matter of physiology. For all animals must use part of their feed energy for maintenance (which for warm-blooded species mostly means keeping warm) and can use only what's left over after that for growth. The amount the animal needs for growth depends largely on the feed conversion ratio - how many kilos of feed are needed per kilo of body weight, which depends largely on the species; and it also depends on the size of the finished animal - or, more precisely, on the difference between the weight at birth and the weight at slaughter. But the amount of energy (and protein) needed for maintenance throughout the animal's life depends in large part on how long it lives. In summary, the quicker an animal can be raised to slaughter weight, the less it will spend on the simple business of staying alive, and the greater the proportion of total energy or protein consumed may be spent on growth.

But there's another complication. In birds and mammals the different body components grow at different rates. At first the emphasis is on the skeleton - so young calves, lambs, foals and puppies are 'all skin and bone', lanky and big-jointed. Livestock killed too young for food are generally very disappointing, and a great waste. But once the skeleton is on its way, the young animal puts on muscle, so young adults are athletic. In general, on a natural diet, body fat builds up only when the basic body form is laid down. The same applies to children in a state of nature: first the skinny kid, then the gawky adolescent (a sudden spurt of growth at puberty), then the young athlete who thinks that he or she will live forever, then comfortable and weighty middle age, and then decrepitude. Childhood obesity is pathological, though now made common by industrial diets.

Farmers, though, must strive to achieve 'finish': animals that are nice and muscular, which is the meat, but also have just the right amount of fat, which makes the meat succulent and provides much or most of the flavour. Modern chickens and turkeys are bred and fed to grow at a prodigious speed - the chickens reaching oven weight in six weeks, and the turkeys practically spherical - and in both species the muscle and fat tend to accumulate well before the skeleton is fully formed so sometimes the benighted creatures can hardly stand (or cannot stand at all). There was a vogue in the 1960s for feeding beef cattle on barley (because oil was very cheap and barley was in surplus) and so they grew fat before they had put on much muscle or bone and hence were 'finished' while still tiny. Then again, when cattle are raised on a more natural diet and at a more natural pace, the fat is far more unsaturated, so is better for the consumer, and is more integrated into the muscle (marbled) which is what really produces succulence. And so on. Farming seeks to modify nature, but it's always best to work to nature's rules.

Yet more complications: the total amount of food required to produce a finished animal must also include a proportion of the mother's feed - for without her, fit and ready to give birth, there would be no baby in the first place. In pigs, the cost of feeding the mother for a year is spread among the 20 or more offspring she produces in the course of it. But a suckler cow produces only one calf per year and typically is not milked either, so her calf, once reared and fattened, must also bear the entire burden of her costs as well. This is why suckler beef doesn't pay unless the cow is fed cheaply, typically on wild pasture. Shepherds generally prefer twin lambs to singletons - but this raises more problems. Ewes that habitually produce twins also tend to produce triplets and since the ewe has only two teats the third lamb has to be fostered or raised by hand which is time-consuming and can be difficult. Shepherds also prefer to know which ewes are likely to produce multiple births because then they can be on stand-by to assist, which they would not normally have to do with singletons. Here another piece of high tech comes in very handy: an ultrasound scanner that can count the number of fetuses in the pregnant ewe with fine accuracy.

Another way to increase the ratio of offspring to mother is to cross a small dam (meaning mother) with a big sire - so small hill ewes are commonly crossed with a big lowland ram to produce fast-growing lambs which are weaned with their mothers and then finished on lowland pastures. The dams themselves are often hybrids and the rams may be too, so the logistics of sheep production can become very complicated. But then again, some crosses work very well and some do not. For example, if small breeds of dairy cattle are crossed with very big beefy bulls, obstetric problems may result, and this is both cruel and costly; and some crosses don't work as well as others, even when the parent breeds seem well-matched. Cruelty results and disaster awaits whenever the output of the animals is raised too far beyond their natural limits. Modern dairy cows, producing at least six times as much milk a wild cow, often fail to reach even a third lactation - though a wild cow may live for 25 years. But still agribusiness clamours for more, and industrial scientists are happy to oblige and are pushing towards the 20,000-litre cow and seeking ways to clone the milkiest ones. Some scientists seem keen to confirm their Dr Strangelove stereotype.

In practice there are many different strategies for raising beef depending on the breed, the time the calves are born, the nature of the pasture and the terrain, the preferences of the market, and the judgment of the farmer. Britain's Pasture-Fed Livestock Association, PFLA, with its ever-growing membership, insists that cattle and sheep are fed *entirely* on grass or browse throughout their lives, but most farmers make at least some use of concentrates. Typically, the young beef animals (steers) are finished indoors on concentrate in winter, or outdoors pasture in summer. So, depending on when they are born (usually either autumn or spring), and how they are finished, they are generally slaughtered at 6 months (for rose veal rather than beef) and sometimes at 12-15 months, or commonly at 18 months, and often at two years plus. On very poor pasture where the land is cheap, cattle may take several years to finish. I know of Highland cattle that are finished at 5 years plus. This is clearly a very low-input, very low-output system - though beef raised slowly on a natural diet should fetch a premium price. Beef raised primarily, or largely, on cereal and raced to the abattoir is not what beef should be.

Enough has been said to introduce at least some of the complexities. I know livestock farmers who have been at it for half a century and are still eager to learn more. Alas, financiers and politicians are still urging farmers and researchers to raise cattle that are more and more productive in enterprises that are more and more elaborate and capital-intensive, with as few human employees as possible. Yet there are some encouraging trends - as we will see in the next chapter.

CHAPTER EIGHT

# Livestock II: Change for the better

A great deal in today's livestock farming is retrograde, going on disastrous, including the continuing obsession with growth rate and fecundity ,and the continuing pressure to raise animals in factories on the biggest possible scale. But many farmers, scientists and food buffs worldwide, are fighting back - making true improvements, not by sweeping aside traditional practice, but by bringing good science and a more refined morality to bear upon it. So here are four good trends:

## The importance of being natural

One cryptic but highly significant advance that applies across the board is a shift of mood - a shift of zeitgeist. Thus in the 1960s (I remember it well) there was a vogue for 'natural' diets - including wholemeal bread, raw vegetables, sea salt and the rest; while those who still wanted meat insisted that cows should be left to graze in flowery meadows while chickens foraged in back gardens. But the 1960s (when I was reading natural sciences at university) was also a time of ultra-hard-nosed rationality: the Enlightenment belief that with enough science and clear thinking we could understand life and the universe absolutely; the reductionist belief that all biology should be reduced to chemistry (and ultimately to physics and maths); and the positivist belief that we should not take anything seriously until we had demonstrated its veracity by experiment and by maths. Nowhere was such hard-nosedness more conspicuous than in medicine and agriculture, the two major practical pursuits that engage most directly with living systems. Human nutritionists rejected all talk of dietary fibre (indeed the term had not been invented then) and most took it to be self-evident that our gut bacteria are of no significance except as hangers-on that sometimes turn nasty. Animal nutritionists had little or no idea of what, if anything, the various herbs that grow in meadows contribute to the wellbeing of grazing animals. It was assumed that cattle need plants purely for the energy and protein they provide - which was best supplied by highly fertilized monocultural ryegrass bred specially for the purpose. Any extra minerals or vitamins could be supplied with custom-made supplements. Since all biology was chemistry, living systems were best left to the chemists. Agriculture was conceived as industrial chemistry al fresco. Biology and the philosophy of science have moved on apace since then but commercial biology has not; and now, more than ever before, commerce rules. The reductionist mentality lives on and the grip of the agrochemical companies with its spin-off of biotech is as powerful as ever. It is, after all, immensely profitable, and profit now is the judge of all things. Yet now it is clear that nature can never be understood exhaustively and that the kind of experiments we are capable of doing cannot reveal all there is to know, and what we call understanding is in truth a narrative that we happen to find convincing, cobbled together from the little bits we do know, or think we know. We need always to be humble in the face of nature. It is naive, and dangerous in the extreme, to assume that the insights we gain from the particular experiments we are able to do at any one time can reveal what lawyers call 'the whole truth'; and, by the same token, to assume that phenomena we cannot explain (and perhaps never will) are therefore not real.

So it is that scientists who think deeply about human nutrition now take the role of fibre and gut microbes very seriously, even though they do not understand either; and they acknowledge the importance of minute and almost unanalyzable components of diet that I call cryptonutrients (Chapter 2); and similarly perceive that the many different grasses and herbs, and indeed woody plants, that grow in 'natural' pasture are probably very significant indeed. Certainly cows on mixed swards run up fewer vet bills. Even more generally, many agricultural scientists now acknowledge the key importance of ecology - the science that seeks to come to terms with the complexity of nature, and with its innate uncertainties. They also perceive, as it was not fashionable to do in the behaviourist 1960s, that animals are not machines. They are emotional beings, and they certainly have some powers of thought. They are mindful, and must be treated as such.

In short, attempts to make husbandry as natural as possible, to plug in to nature's complexity and to respect the psychology of animals, is no longer written off as woolly-minded romanticism - except, that is, by people still stuck in the paradigms of past centuries, whose thoughts are shaped by the promise of profit.

## Welfare

We cannot take it for granted that what we consider to be natural ways of living are always best for an animal, or that the kinds of farming that look most natural are always the most benign. Cages were introduced for chickens in the 1950s partly as a welfare measure - offering freedom from cold, foxes, hawks and (up to a point) flock bullies. I recall, too, when I was involved with London Zoo, one old buffer (a retired colonel if memory serves) regretting the recent demolition of the antelope paddocks - one straw-strewn concrete space per species: 'Splendid quarters for an animal!' But then, the erstwhile colonel and the sometimes well-meaning battery farmers of the 1950s were brought up in the days when animal psychologists and biologists in general were still denying, in the style of the Enlightenment, that non-human animals have any finer feelings at all, or powers of reflection, and seek only to be warm and wellfed.

Somehow we need to strike a balance between the Romantic notion that the warble-struck, half-starved Monarch of the Glen must be enjoying life in the great outdoors, and the hard-nosed 'scientific' dogma which says that animals are just machines that can be happily housed in a box. To make the right judgment, as is true of life in general, we need a combination of compassion, science and intuition - meaning common morality, knowledge and common sense.

First and foremost, we need to give a damn. Many people, I've discovered, don't. Some simply do not accept that animals experience unhappiness, beyond the crude sensations of pain and fear. Others accept that they do but say, 'So what?' Industrial broiler chickens, say, don't live long anyway - just a few weeks - so whatever they suffer, it's soon over. Besides it is still assumed (it's a convenient assumption), that if animals were unhappy they would stop eating and breeding, and since modern livestock out-consume and out-breed their wild ancestors several times over, they must be very happy indeed. But we can infer from human experience that this is simply untrue. Gorging can be a sign of depression.

For those who do accept that animals are sentient and do give a damn, there are three big problems: to decide what, in truth, does make an animal happy; to find ways of judging whether a beast is happy or not; and to devise a system of husbandry that meets its needs but is practicable and does not break the bank. For hobbyists, cost hardly matters. Pigeon lofts and stables have often been prestige architecture. But for those who seek even a modest return, within the context of a social enterprise, whatever is done must be affordable.

Finding out what makes animals happy, and then judging whether they are happy or not, requires a combination of good biology and empathy, reinforced by hands-on experience. Good biology must include physiology (how an animal functions); psychology (what it thinks and feels); natural history/ecology (how the animal lives in the wild and in what kind of social groups); and evolution (what kind of animal is it? What were its ancestors like?). It also requires veterinary knowledge - to see whether the animal is sickening and if so why.

Most of our domestic animals - certainly cattle, chickens and pigs - are primarily woodland creatures and this should immediately tell us that all of them should appreciate a varied diet. Modern cattle are descended from the aurochs which lived in open woodland, and they have inherited their taste for browse as well as grass. Sheep like browse too. The variety helps to keeps them healthy - all those cryptonutrients that just are not present in custom-bred grass. It's now known, too, that animals of all kinds in the wild self-dose: seeking out particular herbs to eat when they feel poorly. So it is that domestic cats nibble at grass now and again; and many a zoo and farm now grows patches of herbs (basically wildflowers) of various kinds within their animals' enclosures. In Cornwall, farmer Ben

Mead applies the same principle to the entire diet of his 130 dairy cows. As he told the first ever Oxford Real Farming Conference in 2010 (see Epilogue), he raises his animals with minimum fuss on natural pasture with plenty of browse - and has found inter alia that they have a wondrous penchant for Japanese knotweed; a good way to get rid of what has become a serious alien weed in Britain. His cattle remain in rude good health. Significantly, though the area is not short of badgers, his cattle have always been free of TB. More broadly, in 2008 his vet bill was down to £13 - for an animal that caught its nose in some barbed wire. From the milk of his cows, Ben and his wife make the now famous Cornish Yarg cheese.

The woodland pasts of cattle and pigs should tell us too that cattle and pigs above all like shade - to keep them in open sun, as is the common fate of modern 'free-range' pigs, is surely cruel; and so too are the romantic cattle drives through the plains of Texas and up to Wyoming as featured in western movies. Chickens like the shade of trees and appreciate the protection they bring from aerial attack, from hawks, crows and gulls. They are often unhappy under open skies. Sheep and goats are less sylvan but they too appreciate shade, as all animals do. Even oryx, supreme desert animals that they are, seek shade in the midday sun if there is any on offer (as I have observed in San Diego wildlife park). All this is another reason for taking agroforestry seriously.

A little natural history tells us that pigs in a state of nature are family animals. The mothers make nests in some secluded spot in the woods when they are due to give birth and then rear their offspring en famille - and the boar often stays in attendance. Adult pigs of the wilder types are formidable and they protect their offspring heroically against all comers. I remember a film of a leopard knocked 10 feet in the air by a wild boar that entered frame left like a steam train as the cat tried to maul one of his offspring. Yet in modern industrial systems the boars are kept well away or, it is thought, they would almost certainly murder their children, while the mothers, not necessarily bred for mothering and with over-large litters, are likely to lie on their babies and crush them unless physically separated from them by an iron bar.

In the late 1980s when I worked for the BBC, I had the privilege of visiting the late Professor David Wood-Gush at Edinburgh University's School of Agriculture, a pioneer ethologist and welfarist, who was showing that things could be very different. He took me to his piggery in the woods where domestic sows were raising their families of piglets in ultra-cheap enclosures made with straw bales, and with plenty more straw for nesting. Sows of a type that retained some mothering instinct raised their babies without harm and it was perfectly safe to leave the boars with them too. For more background see *Managing the Behaviour of Animals*, which Professor Wood-Gush co-authored with Patricia Monaghan.

As woodland omnivores, too, pigs and chickens like to dig and scratch for food - and should be free to do this in captivity, which makes them excellent cultivators. More generally, it is evident from studies in zoos and laboratories that animals like to be in control of their own destiny, just as we all do. They like to choose the details of their diet, and whether to stay in the open air or seek the shade, and be warm or cool. All this is commonsense, which with a little empathy, we would readily surmise would be the case.

But can we really tell whether animals are happy or not? Our perception has come a long way from the erstwhile wisdom which said that animals are mere machines. Way back in the 1980s the Cambridge psychologist Professor Sir Patrick Bateson remarked that we could not possibly understand animals unless, to some extent, we were anthropomorphic; unless we accepted that they may have the same kind of feelings that we do, at least up to a point.

We can go some way to getting inside an animal's mind by measuring stress hormones in the blood, notably corticosteroids. But also, as Marian Stamp Dawkins at Oxford University is showing, we can judge what's going on in their minds by assessing aspects of their behaviour (see What Do Animals Want? a 35-minute video at www.edge.org). We all know that hens in sheds like to lay their eggs in nest boxes - equivalent to the hidden spots they find in the wild at the foot of trees and shrubs. But how strong is their desire to do this? To test this, Professor Dawkins placed balloons that were slowly or quickly inflating, between the hens and the next boxes. Hens are frightened by such oddities. So she was able to gauge just how much trepidation they were prepared to put up with in order to get to the boxes: a measure of their desire to fulfil their natural predilections. Back in Scotland, Françoise Wemelsfelder at the Scottish Agricultural College, Penicuik, has been measuring to what extent farmers or any of us can tell the state of an animal's mind through empathy. She invites people simply to look at any given animal and pass an opinion on whether it is happy or unhappy, depressed or confident. She has found that different people's opinions of the same animal correlate very highly; and that their opinions correlate highly with other more 'objective' measures of stress. Dr Wemelsfelder's work chimes nicely with the idea that commonsense and common sympathy can take us a long way, and with Professor Bateson's observation that controlled anthropomorphism is indispensable.

These are a few general guidelines, some of which are creeping in to modern husbandry, and must play a very big part in enlightened agriculture. Various campaigns for changes in the law to improve animal welfare have scored some notable successes in the past 10 years. But above all we need a change of mindset. We need truly to regard other species as our fellow creatures; or as Robert Burns said to the mouse whom he disturbed when he was ploughing a field: ' ...poor, earth-born companion / An' fellow-mortal!'

Cruelty to animals should not simply be unlawful. It should be unthinkable.

## Grazing

More than half the world's officially designated agricultural land is grassland - one sixth of all land. Britain's farmland is about 60 per cent grass. In its wild state (as in the American prairie) or in its traditional farmed state (as in the classic English meadow - or in the glorious upland meadows of the Alps), this 'grassland' also contains scores or hundreds of other flowering plants, which the Americans call forbs; plus, sometimes, rushes, sedges, ferns and mosses. Well-managed grassland can and often has subsisted and thrived in situ for centuries or even for millions of years, although it may from time to time give way to forest or be drowned in ice as the global climate shifts from hotter or cooler or wetter or drier. The roots of the grasses and other wild herbs may penetrate to huge depths - sometimes five metres or more - and are highly mycorrhizal: the hyphae penetrate the roots and hugely increase their penetrating power and their ability to capture and take up nutrients. Long-established, deep-rooted grass is wonderfully resistant to drought and, thanks to the roots (and the worms), is full of tunnels that help to drain away surplus rain - and so it reduces flooding too.

Many farmers worldwide and through all of history are content simply to leave the grassland in its natural state: unimproved. Domestic sheep and cattle then in effect become substitutes for the wild herbivores of the past. Britain in its pristine state never had vast open spaces but it did have plenty of open woodland, which in the Middle Ages was known as forest and was grazed and browsed by aurochs and deer. North America has long had its open prairie, where the bison (buffalo) roamed, plus pronghorns and deer and so on. But often farmers improve the pasture in various ways, adding lime to counter the acidity (beloved of heather, moss and very coarse grasses) and also phosphorus. Still, though, the vegetation is left undisturbed to form 'permanent grassland' (though it may often be reseeded too).

In addition to permanent pasture - improved or unimproved - there is a place for short-term leys of one, two or three years, when the grass is sown as a standard crop as part of a rotation. As a one-year ley, the grass is of course an annual. Grass for short-term leys may be monocultural ryegrass, bred to be high in sugars and especially in protein (when highly fertilized). Nowadays more sophisticated mixtures of grasses are generally preferred - some very nutritious, some especially hardy, some able to withstand trampling; with clover and other legumes mixed in to fix nitrogen. There is certainly a place for short-term leys but they can never acquire the deep roots and the copious mycorrhizae that give such resistance to the caprices of the weather, or produce the kind of soil structure that improves drainage and removes panning - and contain the variety of herbs with all its cryptonutrients. Very highly fertilized, more or less monocultural, leys can give the most energy and protein per unit area, and produce the fastest growth and the highest yields. But animals raised at more leisurely pace on varied, natural pasture are superior in flavour to those that are intensively fed (on concentrate or on custom-bred monocultural grass) and raced to slaughter and, as noted in Chapter 2, they seem to be nutritionally superior too; and in meat, flavour and nutritional details matter far more than mere bulk.

Alas, much of the world's permanent pasture is very badly managed. Often it is undergrazed, so the grass becomes rank and woody plants start to take over. More often it is overgrazed, so that the vegetation dies and the soil is laid bare and cracks and is blown or washed away. Meanwhile the grazing cattle (and sheep) belch out copious quantities of methane and can be shown at least statistically to contribute to global warming. All this has helped to give pastoralism in particular and livestock in general a bad name - and has also allowed opportunist industrial farmers to argue that cattle should be more or less permanently housed and fed largely on concentrate, when the methane they produce can be filtered out, and never escape to the atmosphere at large.

But various pioneer farmers worldwide have shown in recent years that permanent grassland can quite easily be managed in ways that keep the vegetation intact and year by year improve the soil beneath. Allan Savory of the Savory Institute promotes what he calls 'holistic management' of farms, grassland and regions; and Joel Salatin puts the ideas into practice on his farm in Virginia. The core idea is that of 'mob grazing'; a radical break with traditional grazing strategies. Traditionally, farmers release their cattle or lowland sheep on to pasture for a few weeks or months in the summer, or leave them out all year if it's not too cold and the ground is not too muddy, and let them get on with it. But cattle (or sheep) left simply to wander in a field never work the pasture to maximum efficiency no matter how finely the stocking rate (animals per unit area of land) is adjusted. They simply pick the best bits on any one day and then wander on to the next best bit. They leave their dung in the field which replenishes it but they also breathe out copious quantities of carbon dioxide in the normal course of breathing, and belch out methane from the rumen.

Mob grazing is quite different: much more tightly controlled - but also, albeit paradoxically, closer to what grazing animals do in the wild. First, all the animals are gathered into a small space, generally circumscribed by an electric fence. They are close-packed; as many as 50 animals to an acre. Between them they eat the grass right back, from long to short (though not too short!), and they eat it evenly. Then after 12 or 24 hours the whole 'mob' is moved along to the next patch - and so on all the way round the grazing area. After several weeks or months, and usually at least six weeks, they get back to the patch they grazed first. By that time the grass on the first grazed patch has recovered.

The mob of cattle in their restricted space don't eat so much that the soil is laid bare, or enough to kill the grass, but they do reduce the leaves very significantly. The roots beneath depend for their nourishment on the leaves, and so are profoundly affected. Exactly how they are affected doesn't seem entirely clear at present. Some say the roots die back but this does not appear to be the case. Instead, it seems, the roots exude large quantities of carbohydrates which encourages huge populations of soil microbes to build up around them. Either way, soil carbon is built up - either by dead roots which then decay (slowly because they are full of mycorrhizal hyphae) or (more probably) through the exuded carbohydrate and the microbes that come to feed on it. Whatever the mechanisms, trials (though not yet enough of them) have shown that by this means the organic content of the soil can be built up remarkably quickly. This has all kinds of advantages as discussed in Chapter 5. But also, it seems, the amount of carbon sequestered by well-managed mob grazing exceeds the amount that the animals expire in the form of methane and carbon dioxide. This means that the net effect of animals grazed in mobs is not to raise atmospheric carbon, but to reduce it; and since the soil is able to hold far more carbon than the atmosphere does, the effect of this could be very significant.

In other words, just to rub the point home, the modern, fashionable idea that it is better for the climate to take cattle and sheep off the land and feed them indoors on cereal and soya (or on grass harvested from the surrounding fields and brought to them) is wrong. Yet we should not be surprised by this. A recurrent message throughout this book is that most of what has become accepted agricultural lore is wrong. This is not surprising, because agricultural strategy these days is not framed by bona fide biologists including ecologists who truly desire to know how nature works. Still less is it framed by hands-on farmers (as opposed to agribusiness people). It is framed by economists and business people, supported by politicians and their hand-picked advisers, all of them committed to the neoliberal idea that the point of farming, as of all human endeavour, is to maximize short-term wealth. In the short term, as things are, with oil still affordable and with cheap foreign labour, it is most economical to industrialize all farming - which includes raising cattle indoors on imported feed or chemically enhanced bussed-in grass, with filters to extract the methane. The ecological approach to farming in general is largely neglected in official circles (the kind that soak up tax-payers' money). It is left to private individuals to develop.

How does all this square with the idea that husbandry should imitate nature? Is it natural to pack animals together so tightly, and marshal them round a field with an electric fence? More natural than it seems, is the answer. For herds of animals in the wild, at least in open spaces as on the prairie or the savannah, move around in tight packs as a protection against predators. If they drifted around without a thought in their heads, like Jerseys or Ayrshires in a traditional meadow, they would be picked off like ducks at a fair. The electric fence, in short, is a stand-in for the lions and hyaenas of Africa or the wolves in North America that would otherwise be eyeing them up.

Once the mob-grazing regime is set up, other enterprises can be tacked on to it (as discussed in Chapter 12).

One last, whimsical, evolutionary note. Grassland has never been more extensive than in the period that lasted from the mid Miocene, say 15 million years ago, to the late Pleistocene, about 100,000 years ago. The grasslands were grazed through all that time by billions of cattle, antelope, pronghorns, sheep, deer, horses, elephants, rhinos and camels, of many different species. If it's true that grazing herbivores cause global warming the world should by now be as tropical as it was in the Eocene, some millions of years before the Miocene. Instead, with a few ups and downs, the world grew cooler through the Miocene and the following Pliocene and we finished up with the Pleistocene ice ages. To be sure, scientists tell us that the ice ages were caused primarily by the vagaries of the cosmos - the

Earth's varying distance from the sun. But this would not lead to ice ages unless the world was already cool - which, clearly, billions of grazing animals, moving in mobs, did not prevent.

## Small is indeed beautiful: the micro-dairy

In this age of industrial agriculture, big is assumed to be best because it supposedly brings 'economies of scale'. But there is nothing in farming that cannot be done on the small scale. If we shift the criteria of excellence away from mere quantity and short-term profit, to quality of food and of life; if we take the damage of big-time farming seriously; if we change the economic rules, so the subsidies and tax breaks are not so assiduously aimed at those who are already big and rich; and if we replace the ruling oligarchy with governance that is on the side of humanity and of the biosphere - then small-scale farming is almost always better. As the economist-philosopher E F 'Fritz' Schumacher remarked (it is the title of his most famous book): "Small is beautiful".

Some small farmers are part-timers; their farms are small, or else they focus on just one enterprise, so that they have time for other things; music, accountancy, teaching, science, medicine, carpentry, whatever it may be. Some are full-time smallholders, often working (in due season) all the hours that God made. Some are hobby farmers - who are often, but wrongly, despised because area for area they can be among the most productive, and worldwide they make a huge contribution and certainly help to raise the quality. Hobby farming for many is just an indulgence - but so what? A person who spends a fortune on the avian equivalent of a five-star hotel for his or her half-dozen chickens - well: what else would they spend their money on? If the chickens are bona fide Dutch Welsummers they are helping to perpetuate one of the world's finest breeds which lay some of the world's finest eggs and are great foragers on snails and weeds and what you will. The hobby farm too offers a well-trodden path into full-time farming - and we need at least eight times as many farmers as we have now.

We'll discuss all this in Chapter 12, but here are a few salients.

Very much against the commercial-government tide, but very much in the world's interests on all counts, is the renewed interest in micro-dairies. Nick Snelgar's Maple Field Milk is a fine example (pages 146-7).

## Back to mixed farming

We see livestock at their most advantageous when they are a part of a polyculture - key players in the farm-qua-ecosystem. Mixed farms, with some arable, a small dairy, some sheep, pigs and chickens, and at least some horticulture, all as integrated as possible, were probably at their height in Britain in the 1950s. The greatest examples of all are probably those of Southeast Asia, with (as I have seen first-hand) carp and ducks in the paddy fields, horticulture and small plantations on the higher ground, pigs and chickens in the villages, and water buffaloes to do the work and supply milk and meat. All farms too, these days, should be conceived as exercises in agroforestry, as discussed in Chapter 10. Ideally, farms would be run as co-operatives (there are many variants) and as far as possible integrated into their local communities.

Such mixed, co-operative, community-orientated farming is obviously good for humanity at large in many different ways and is an exercise in sound biology. Animals in such systems are certainly not a burden. They become our indispensable allies.

Hornton Grounds Farm (page 148) is a good example of mixed farming.

## Maple Field Milk: a micro-dairy in Hampshire

Neoliberal-industrial thinking favours scale-up, which for dairy means herds of many hundreds of animals – even up to 30,000. The animals are permanently housed, and each one gives at least 5,000 and preferably nearer 10,000 litres (2,000 gallons) per year. The cows are hugely stressed, they don't live long (three lactations at most) and they must be fed on highly concentrated diets including lashings of grain and soya which must be bought in at huge cost, much of it from halfway across the world. Despite all this, if all goes well, the industrial route can be hugely profitable. But it is also hugely precarious. The investment is enormous yet commercial success depends entirely on the global market which in turn is yoked to the ups and downs of oil. For reasons both commercial and biological such 'mega-dairies' surely cannot represent the long-term future. The NFU has recently warned that if we continue to put our faith in scale-up there could be fewer than 5,000 dairy farmers left in the UK by 2025 (down from 36,000 in 1995).

The antidote is the micro-dairy – such as Nick Snelgar's Maple Field Milk in Hampshire. He aims to run 18 Ayrshires entirely on grass (he owns a little land and will rent the rest), each producing around 4,000 litres (800 gallons) per year. Conventional wisdom has it that the income must be far too small – but it doesn't have to be. For Nick does not sell en masse to a supermarket chain at 20-30p per litre, but to local shops at nearer 60p per litre – the same price as mass-produced milk. So customers can then choose between mass-produced, homogenized milk from anonymous sources, and milk from local, pasture-fed cows, fully traceable and delivered fresh. For most, it's no contest. Nick is selling directly to customers at closer to £1 a itre. He reckons the optimum number of households to be 400.

Nick's next step is to build his own dairy herd: he's looking for people to rent a cow. He's built his own two-birth mobile milking bale of the kind first popularized in the 1920s, which can be pulled to where the cows are grazing by a tractor or a small horse or even, if necessary, by human draught power. This saves on some of the standard fixed costs, such as hard standing and permanent farm tracks. He then adds value by pasteurizing the milk on site – at the moment he is buying milk from a local dairy farmer at 40p a litre and selling it fresh to 13 local shops. At the time of writing the enterprise has been going for 18 months and is 'washing its face'; not fully operational, not making a fortune, but paying its way. New businesses can rarely hope for more.

As Nick and others elsewhere are showing, micro-dairying offers a low-cost route into farming. It is sound ecologically, morally and socially, and offers a long-term future as the high-tech, high-capital, high-input mega-units surely do not. Nick now aims to help establish five similar micro-dairies around the country. Micro-dairying should become the norm as once it was.

Maple Field Milk is conceived as a social enterprise with the legal structure of a community interest company (see Chapter 13). The group that I am involved with, FEA (Funding Enlightened Agriculture – see Resources), is proud to have helped to get MFF up and running.

## Hornton Grounds Farm

Catherine and Graham Vint both farmed in Jersey, then moved out of farming for a while (she into marketing, he a policeman). Then, they bought the beautiful Hornton Grounds Farm in the English Cotswolds on the Oxford-Warwickshire border. The soil is 'Cotswold brash' (very stony) and had been farmed conventionally for arable. It has taken Catherine, Graham and their business partner Simon Thomas five years of prodigious toil to bring the soil and the landscape roughly to where they want it to be. Good farming can't be hurried. Now Hornton functions as a privately owned mixed farm run by Thomas and Vint Ltd.

The secrets of their success, in two words, are quality and diversity. Catherine is a leading member of the Pasture-Fed Livestock Association, the PFLA, and the couple raise prime Aberdeen Angus and Texel x North Country lambs entirely on pasture. The animals graze in summer and feed on lucerne-rich silage in winter. Their poultry and pigs, hardy Tamworths raised outdoors, are fed on home-grown wheat or barley, plus the lucerne-rich silage. Straw from the cereals provides the animals' bedding and the resulting muck is spread on the fields. In essence, it's all very traditional. The meat is butchered at a local abattoir – the beef then hung for three weeks – and sold mainly through a local butcher.

But that's not the end of it. Hornton also offers four rooms for B&B, and the Cotswold barns have been converted into modern stables which they offer for livery. They have planted a mixed wood by the lake which they hope will add yet more wildlife to the hares, foxes, deer, skylarks, buzzards, lapwings and miscellaneous passerines which already abound. There are bridle paths for horses, bikes and walkers and a croquet lawn and a tennis court for good measure. Who wouldn't want to stay there?

The Vints and Thomas are demonstrating that a middle-sized mixed farm that maintains the highest standards of agroecology can succeed, even though within the present economy it is, perhaps, the most endangered kind of all. All it needs is expertise, a very good business sense, an excellent location, and astonishingly hard work.

CHAPTER NINE

# Horticulture

Horticulture is surely the most ancient form of agriculture - it can be said to have begun when our long-distant ancestors first started protecting their favourite food plants, which a great variety of animals do in one way or another. Yet its best days might still be ahead of it. It does not and need not occupy anything like as much land as arable farming or livestock (Britain's officially recognized horticulture occupies only about 145,000 hectares out of a total agricultural area of 18.2 million hectares) and in some countries, including Britain, it tends to be seen as a commercial also-ran. But horticulture is surely the key to the agrarian renaissance, at once a lead player and a fifth columnist. It offers the principal route by which people at large can become involved in farming, so that enlightened agriculture and all that goes with it, ecologically, socially, and politically, can become the norm. Horticulture at the highest level requires tremendous skill, boundless knowledge and absolute dedication, but anyone can dip their toe in the water - and once dipped, on patio or rooftop or window-box, who knows?

There's a rich literature on horticulture of all kinds and at all levels (see Resources) so I can't hope to add significantly to what's out there, even if I had the expertise to do so. But a few broad points are worth bringing out.

## The variety of horticulture

Horticulture is often practised these days on vast estates but it's particularly suited to the small scale. Skilled amateurs even in northern climes like ours may keep their entire families in vegetables through most of the year on 10 square metres (a kitchen-sized patio) while others make a good living from two or three hectares. Cities are great for horticulture - especially now that lead, once the universal urban pollutant, is banned from petrol. Buildings act as storage heaters and suntraps, and it's amazing what can be grown up walls - the vertical garden - and on rooftops; where the soil and vegetation help to insulate the building.

Somewhat confusingly, there are distinct schools of horticulture and, mysteriously, all at their best seem to work extremely well, with total yields per unit area far above what can be achieved with arable or livestock. Thus in the decades after World War II, when I first started taking an interest, growing was seen as a heroically physical pursuit. In and after the war the government encouraged my father's generation to plant vegetables and the illustrated manuals and newspaper strips (it was a great age of graphic art) invariably advised their readers to double dig. The top spit, the topsoil, was reserved; then a second spit was taken out and as much manure or compost as possible was thrown into the trench, then the earth put back with the topsoil at the top as before.

In truth, it you don't know what's underneath your plot when you start, then such deep cultivation is well worthwhile, not to say de rigueur. When I started a vegetable patch in our south London garden in the 1970s I went down four inches and struck the brick floor of some erstwhile farm building with (it turned out) a skip-full of hardcore beneath. A relative of mine in his new garden in north Kent unearthed an entire motorbike. But once you have cleared such hazards, the joys of deep or double digging can be hugely overstated. The point of double digging (apart from removing hazards) is to mix manure or compost as thoroughly as possible with the soil. But on a scale of an acre or more this can be achieved more easily with a small plough, which might even be pulled by a small horse or two - as demonstrated not least by Ed Hamer on his three-acre market garden at Chagford on Dartmoor (see below). But cultivation in general should be minimalist, for more than just enough can do far more harm than good. Too much vigour disrupts the worms, the great aerators, drainers and mixers, and other small creatures - and the mycorrhizae; and can bring death and decay to the soil flora and fauna, and expose organic matter to oxidation, so releasing soil carbon.

Horticulture of the kind recommended during and immediately after World War II tended to be organic of necessity. The agrochemical industry was not yet fully into its stride and horses still abounded even in cities, so growers who were quick with the shovel could fertilize their gardens for free, delivered if not to the front door then at least to the road outside. Gardening of the kind now encouraged by many a garden centre is often an exercise in commercial chemistry and packaging. In contrast to both, enlightened growing should be as organic as possible as a matter of strategy. But beneath the broad umbrella of organic growing are at least two approaches with distinctive philosophies over and above the general fact of being organic.

The first is permaculture. The idea of it was put in place by the One-Straw farmer Masanobu Fukuoka and the term permaculture was coined in 1978 by the Australians Bill Mollison and David Holmgren. It seeks to create a productive ecosystem that continues with as little interruption as possible throughout the seasons and year after year, giving crop after crop with minimal and preferably with zero cultivation - as in a wild ecosystem. So it makes as much use as possible of perennial plants, both woody and non-woody: the forest garden is permaculture taken to a logical conclusion. There is maximal mixing of crops, too, to take advantage of synergies between them and to make life as hard as possible for pests. Fertility is added in organic form - compost - from the top, primarily as mulch, which the teeming worms (up to 25 per cubic foot of soil is normal) obligingly mix in. In some systems small plants are planted *through* the mulch.

More romantic (for want of a better word, though actually it is quite apt) is biodynamics, initiated by the Austrian philosopher Rudolf Steiner in the early 1920s. Farming and growing are integrated into a whole philosophy of life and indeed a whole cosmology. Farming practice is heavily influenced by ideas of a spiritual nature, with astrological connotations. For example, planting and harvesting are timed almost hour by hour in part according to the calendar, not least to the phases of the moon; and one of the recommended practices is to bury a cow's horn packed with ground quartz to capture cosmic influences. Steiner insisted, however, that he derived his ideas from traditional practices and that they should be put to the test - and one of the farmers who does just that is Ed Hamer. He says he finds biodynamic practice most effective - particularly planting according to the calendar. It seems to work even though he acknowledges that neither he nor anyone else knows *how* it works (though the moon's gravitational effect on the movement of Earth's water is a possible clue). But he seeks as far as is possible on a commercial holding to try out the methods on a controlled, plot-by-plot basis, some cultivated biodynamically and some not.

Hard-nosed scientists tend not to feel comfortable with any idea or practice that they cannot rationalize exhaustively, and many hard-nosed farmers feel the same. But farming in the end is craft, and crafts can never be explained exhaustively, for explanations are narratives and no narrative can ever be complete. There are always unknowns and unknowables. Ed Hamer, who can think as clearly as anybody, insists too as many farmers do that it's important to respect the ancient roots of farming, and to ask what their predecessors did, and why - and biodynamics is rooted in ancient practices. More generally, the point of all ritual (if we think that biodynamic practices are ritualistic) is to focus the mind. Focus leads to tender loving care and in horticulture as in all husbandry, TLC is the sine qua non, although it all too often goes by the board these days, replaced by chemistry and graphs and balance sheets. In short, the success of biodynamics is hard to explain but many excellent growers swear by it, and excellence must be the aim.

Otherwise, organic gardening follows the general rules of agroecology: maximum diversity; intercropping; trying always to avoid bare soil, with mulching and green manures; adding fertility with compost and by nitrogen fixation, courtesy of legumes; keeping all the soil nutrients within the physiological range that the crops prefer; seeking to maximize soil carbon - and the living components of the soil, microbial and otherwise; and by crop rotation. The latter is key, potentially complex, and deserves a separate note.

## Crop rotation

Rotations vary in detail for all kinds of reasons but the general principles are the same always, whether in arable or in horticulture. First, successive crops should be botanically dissimilar, unrelated, so that pests and diseases cannot easily hop from one crop to the next. Secondly, the sequence should begin with high fertility and with crops that favour high fertility. The next crop in the sequence should be of a kind that favours less fertility, and so on; although the grower generally boosts the fertility at intervals, sometimes with more manure or compost and *always* by including a leguminous crop that fixes nitrogen. Finally, there should be a year or two in the sequence devoted to green manure or grazing, when no crop is taken off.

Ed Hamer favours an eight-course rotation. This means in practice that the whole plot is divided into strips, and each strip is divided cross-wise into eight sections. The sections in effect represent the successive years; with each new year, each crop in each section in each strip, shuffled to the next section down.

So:

- First, section one in strip one is manured - to be followed immediately by potatoes. Potatoes are hungry plants (they really appreciate the manure) and are deep-rooting, breaking up the soil and creating channels through it as they grow, with further breaking up as they are harvested.
- In the second year, the potatoes are shuffled to section two while section one is devoted to brassicas - also hungry, but not as much as potatoes. After the brassicas Ed adds more manure.
- Brassicas are followed by cucurbits - courgettes and squashes.
- In the fourth year comes a crop of legumes - peas, beans, lentils.
- After the legumes come the umbellifers (though the traditional family, Umbelliferae, is now re-named the Apiaceae). These include celery, celeriac, carrots and parsnips.
- Year six is devoted to alliums: onions, leeks, garlic.

Then come two years - meaning two plots - of fallow, alias green manure: the first plot with crimson clover, white clover and rye; the second with clovers and rye grass. After the second year of fallow it's time for a heavy dose of manure then back to the potatoes again.

This whole process may seem complicated but the logic and the basic sequence are clear enough: potatoes, brassicas, cucurbits, legumes, alliums, umbellifers (parsnips, carrots), green manure, then more green manure.

All this can be done with a spade or an adze and in general, the simpler the better - but spades and adzes are very hard work and there is no great virtue in being a slave. Two million years of tool-making experience have shown that appropriate technologies really can make all the difference.

## Appropriate technology: tools for conviviality

If you are farming on a vast scale then you are obliged to use vast machines or you can never hope to cultivate or harvest within the brief windows of time when these things need to be done. If you just have a window box or a patio then all you really need is a trowel and a dibber, and a watering can. But if you are cultivating just a few acres as the typical market gardener does then you have a wide choice. You could do most of the work with ordinary garden tools - spade, fork, hoe, rake - if you had the energy. Or you could scale up to a cultivator - one you push or one attached to a quad bike. You might even invest in a small tractor, if you could find one. Or, like Ed Hamer, you could make use of small horses.

All approaches have their advantages and drawbacks. So how do you decide what's most appropriate? As a rule of thumb, it pays to be minimalist: don't sink capital in fancy stuff you don't really need and depreciates by the month. Yet I always remember the advice a builder once gave me when I was wondering whether to invest in a new word processor: 'Never economise on tools that really can make your life significantly easier!' Key too is a lesson we can learn from Britain's stately homes, from the days when they were also working farms - that the holding is the workplace where you spend most of your time and should be made as agreeable as possible and indeed, as William Morris emphasized, should be beautiful. Many industrial farms suffer, and so do the people who work on them, because the environment is so unpleasant.

Practical people and philosophers alike have wrestled with the issue of appropriateness over many centuries. Francis Bacon at the start of the seventeenth century said in effect that science had the potential to make us omniscient (though he remained a godly man) and technology would make us omnipotent, or as near as makes no difference. So he helped to kick-start the Enlightenment. Many raised doubts along the way, including the much maligned and misunderstood Luddites in the early nineteenth century who broke the various textile machines that were putting them out of work. But on the whole the tide has gone with Bacon. Many seem still to believe that with more and more science and tech we really *can* understand everything, and do everything we might want to do; and this view has prevailed not because it is correct, but because in practice it has brought great wealth to a few, and the wealthy few dominate and set the general tone. The twentieth century also revealed the lie of this idea; showing (as briefly outlined in Chapter 4) that science can only do what science does, and must always leave us far short of omniscience; and demonstrating in spades that if technology is driven simply by a mindset that equates progress with power, wealth and dominance, then science and high tech, which should be among our greatest achievements, all too easily become our greatest threat.

As things stand, we *do* have great knowledge, albeit far short of omniscience, and a tremendous range of technologies that enable us to work what our forebears would have thought were miracles. But while some of the miracles undoubtedly are good (who could question the value of vaccines or modern obstetrics?), some are specifically designed to do maximum damage (more bang to the dollar as the arms traders say), while many that promise to do good in practice cause tremendous collateral damage. So how can we steer a sensible course? How do we identify and promote the kind of research and the technologies that really can make the world a better place, and leave the rest to moulder? This has become the key dilemma of humankind.

For my money, the thinkers and doers who did most to answer that question in the twentieth century are the American architect Buckminster Fuller, the Austrian social philosopher and Roman Catholic priest Ivan Illich and the German-born economist E F Schumacher. Buckminster Fuller argued in a general way that technologies including those of the most ingenious kind must above all serve the best interests of humankind. He is best known for the geodesic dome, made from alternating hexagons and pentagons (the material doesn't matter but glass is good) that interlock to create structures that can enclose huge spaces economically and in ways that are aesthetically pleasing.

Illich, among many other achievements, wrote the wonderful *Tools for Conviviality* in 1973. For him conviviality largely meant autonomy: control over our own lives. Some technologies, he said, bring us greater freedom to do our own thing - like the bicycle and the telephone. But some technologies tend primarily to reinforce top-down government and tend ultimately to enslave us. He put public broadcasting in that category. In those days, after all, there were only a few radio stations, like the BBC, and they spoke to us *de haut en bas.*

In *Small is Beautiful*, also written in 1973, Schumacher argued that technologies should, above all, be *appropriate*. Often the real needs of human beings are best served by small-scale or medium-scale technologies: and sometimes old-style, traditional technologies do the job better than their modern replacements.

Like the Luddites, both Illich and Schumacher have been much misunderstood. Schumacher's phrase 'small is beautiful' has often been taken to mean that small and homely is *always* best. The suggestion that wooden carts and windmills may do just what's needed at very low cost is sometimes taken to mean that petrol engines and electricity have no role to play. But neither of these two sages said what people often think they said. The telephones that Illich favoured *are* high tech, since they are the fruits of formal science, rather than of craft; and so too are bicycles that are truly functional, with their vulcanized pneumatic tyres and a host of fancy alloys.

In horticulture, and small farms in general, we see the principles that Fuller, Illich and Schumacher espoused laid out before us - as evident indeed on Ed Hamer's holding. He cultivates the ground with the aid of two ponies - the traditional way; but the ponies pull devices for harrowing and seeding and the rest which look very simple and old-fashioned but are made, in the US, from light, strong and therefore high-tech metals, and designed very cunningly and probably with the aid of computer modelling to carry out their various functions as efficiently as may be managed.

Similarly, John Letts, the archaeobotanist turned thatcher and arable farmer who featured in Chapter 6, and Simon Fairlie in Somerset, editor of *The Land*, both advocate small-scale arable - and both are aficionados of the scythe. But the scythes they use are of Austrian steel, a triumph of modern metallurgy. Or then again, traditional farmers the world over may reject GMOs for all kinds of legitimate reasons and so are said to be superstitious and backward and ignorant and anti-progress - yet they prove that they are none of these things by making full use of ultra-high-tech mobile phones, and mending their own motorbikes by the side of the road. In truth they are far more sophisticated than the experts who are pushing the latest gizmo. They take what high tech they need and reject what they don't. This is the exactly what's needed. Similarly, a small farmer in Britain may find that the most appropriate buildings might be made from daub and wattle (I hesitate to say who is doing what and where, lest the planners close in with their bulldozers); but nowadays there is a host of light, strong, materials made from renewable resources, including hemp and sawdust, and often conveniently pre-fabricated, that enable almost anyone to custom build their own sheds and houses in a few days, at minimum cost, all of it easily dismantled and repositioned, just as needed. So, too, one of the most convivial technologies for the modern horticulturalist is polythene and a range of other plastics with various improvements. The polytunnel is invaluable, enabling growers in temperate lands to raise crops they otherwise could not, and extending the growing season by several weeks - a big deal in countries where the growing year is short. Polythene tubing makes micro-irrigation cheap and easy; and judiciously positioned sensors and microchips make it very precise and efficient.

More broadly, true modernity does not mean what naive politicians and gung-ho scientists and cynical commercial opportunists think it means. Even the highest of high technologies have their place but they do not in themselves represent progress. Truly modern farms or market gardens are a judicious mix of the ancient and ultramodern: mixed farms with traditional breeds and sometimes with traditional varieties or even landraces, all very quaint and nostalgic to the untutored eye; sometimes with buildings that might have been put up 5,000 years ago; yet all of it artfully underpinned and supplemented by science and technologies that are truly appropriate - which might mean scythes and willow hurdles but might equally well mean electric fences and microchips and mobile phones and polythene. It is all a matter of knowing what you are trying to do and what is truly worth doing, with knowledge of the present and respect for the past and the genius of our forebears.

It's clear, though, that if you get it right, then horticulture can be among the most fulfilling of all occupations. It needs skill and knowledge (never too much), good science, a good state of mind, good relations with the community at large - all human qualities are brought to bear. Already, worldwide, as for the past 10,000 years, many millions (surely billions) of people have been showing what can be done. Yet it's possible that the best is yet to come.

## Horticulture as the key to the future

The vegetables and fruits provided by horticulture are of course essential. They provide nutrients in the form of vitamins, minerals, and (surely) a great range of cryptonutrients, and diverse forms of dietary fibre; each (doubtless) with its own special qualities. They also provide flavour, texture and succulence - able without further input to turn worthy but dull staples into fine cooking, of the kind that people actually like to eat.

But horticulture - at least as practised in Britain and in much of the 'developed' world - does not produce staple crops in significant amounts, apart (sometimes) for potatoes. Even the pulse crops grown in gardens and market gardens are generally intended to be eaten fresh - not dried as grains. Some sunflowers and mushrooms may be grown, but not generally in great amounts. Western-style horticulture does not produce significant quantities of macronutrients, the energy foods (including fats) and protein that we need in reasonable bulk day by day. So it tends in people's minds, including the minds of lawmakers and administrators, to be relegated. Serious agriculture is perceived to mean arable and large-scale livestock - the latter either feeding on grass, as cattle and sheep should do, or else fed largely on grain and perceived as offshoots of arable (which in turn is perceived as field exercise in industrial chemistry). Commonly, too, these days horticulture is practised on the industrial scale, which can be impressive and perhaps has its place, but also has its drawbacks as all industrialized farming does. It is not the only option.

It would be good to redress some of the balance - as indeed around the world many growers are already doing. As outlined in Chapter 6, cereals and other grains can be grown on the horticultural scale, though this doesn't become truly worthwhile unless growers break free of the standard commercial shortlist of acceptable varieties and start to resurrect the many traditional varieties that are now being pushed aside. It can be hard to obtain the seed because it is against the law to trade in seeds that are not on the approved list; but this is one of many laws besetting agriculture that have been made without sufficient thought and under pressure from various lobbies and deserve to be challenged. Pulses, too; beans of the genus *Phaseolus*, in particular, (generically known as kidney beans) come in a vast range of colours, patterns, sizes, habits of growth (climbers, dwarfs, etc) and flavours. A great many, if not all of them, deserve to be kept available for general use. Heaven forfend that beans should *literally* mean Heinz (a small form of *Phaseolus vulgaris* in tomato sauce) and nothing else. A grower with two acres, which is enough for most people to manage, could accommodate cereals by adding another two acres, which need very little work for most of the year. Turn the result into specialist flour and set up a small bakery, and sell the straw for thatching (old-style varieties of wheat have long, strong stems) and it can be commercially worthwhile (and fun, too).

Horticulture should always benefit from some livestock, and then the market garden becomes a smallholding. Chickens integrate easily into the smallest gardens, with some supplementary feed, and larger-scale market gardens can produce enough spare material for pigs. Some object to all animal husbandry as a matter of principle, but others claim simply that animals are inefficient. We outlined all this in Chapters 7 and 8. Suffice here to cite the principle of *land equivalent ratio*, of which more in the next chapter. In this context this means that the added livestock might indeed reduce the output of vegetables somewhat, since some crops might be grown specifically for the livestock. But the total food value of the livestock and vegetables together is greater than the food value that would be obtained if the whole area was devoted to vegetables. In short, if you get the ratios right then adding livestock brings net gain. You could add an acre or two of grazing or fodder crops to the market garden without significantly increasing the amount of work; and if this is possible, why not?

If we add both livestock and small-scale staple crops to the market garden, we have a small mixed farm. The fields put down to grass or fodder crops or to arable can be integrated with the horticulture as a whole into one grand rotation. Then we truly have agroecology - the farm qua ecosystem. On a different tack, horticulture has wondrous social connotations. Many a collection of allotments has turned itself into a club and become a key component of the community - often creating a community out of what had been a mere aggregation of people. If the community, however it is formed, then starts to run the erstwhile assemblage of allotments as one big integrated unit, we have a true community-run farm. Alternatively, as demonstrated in particular by Joel Salatin in the US, existing commercial farmers may invite newcomers to set up small enterprises on their own farms, and so turn relatively simple units employing just a few people into complex systems that may employ many more, not as wage slaves tied to a single strategy but as allies contributing voluntarily to a greater whole. Here we see the principle of the land equivalent ratio in full flight - complexity leading to huge advantages in biological efficiency and in social value. More in Part III.

In short: in addition to all its other advantages, horticulture probably offers the most realistic route in to farming for most of the new generation of farmers that Britain and most of the rich world now desperately needs, even if the powers that be don't realize this; and horticulture also offers the principal route whereby growers and farmers working alone can turn themselves into functional communities that are also commercially viable - which is what commercial companies ought to be.

Here are some examples:

## Chagfood Community Market Garden

Ed Hamer runs Chagfood, a six-acre market garden on organic certified land at Rushford Mill Farm near Chagford on Dartmoor in Devon, along with his business partner Nicky and two tough little ponies, Samson and Billberry. These animals, with ever-increasing skill, pull a variety of ingenious small-scale, lightweight, cultivating machinery, largely imported from the US. Every week from July to February the team provides boxes or bags of fresh vegetables (picked on the day of delivery) for their loyal circle of local buyers. Two hundred is the target and they seem well on their way to it.

That at least is the bones of Chagfood – but their evident success depends on more. First, although Ed is young (around 30) and a first-generation farmer (his parents were teachers), he trained at one of the world's most renowned schools of agriculture at Aberystwyth in North Wales, and is definitely expert. He is also a local Dartmoor man with a strong social awareness who truly believes that a farm should be a community resource. So locals get involved: Chagfood is an example of a CSA, community-supported agriculture (a partnership between farmers and the local community). The members pay upfront, by the year or by the month, by direct debit. In 2015 Chagfood offered three sizes of veg share: large, for a family, costing £660 for the year or £55 per month; an intermediate share at £500 per year or £41.66 per month; or small, at £360 per year or £30 per month for individuals.

'It's very important to establish the customer base before you even start,' says Ed. 'Don't grow the stuff and then look round for someone to sell it to.'

He spends much of the winter keeping people on board (some drop out each year – but more join in). The in-season veg – some of it unusual – comes with culinary advice. Add-ons are important, too, 'value adding'. Chagfood runs open days; people love to see the horses working. They are invited to lend a hand, too: a Chagfood poster/flier in the summer of 2015 read:

'Annual potato-weeding party tomorrow (Friday 12th July) [with] tea & cake and join us for a swim in the river. It would be fantastic to have as much help as possible – even for an hour or so.'

In all, over the year, their volunteers have contributed the equivalent of 240 full working days.

## Organiclea: a workers' co-operative

The river Lea flows south-west from Luton to join the Thames around Waltham Cross – and its banks, once packed with market gardens and smallholdings, served until recent decades as the 'breadbasket of London'. Then, inevitably, the Lea Valley was urbanized – but since 2001, Organiclea is returning at least a part of the valley to what should surely be seen as its natural role. I was introduced to the project by Hannah Claxton, and it is rapidly becoming one of my favourite places, ticking all the right boxes.

First, on its four-and-a-bit hectares (12 acres) it raises a marvellous variety of vegetables, saladings and fruits – including an impressive vineyard, though this has been somewhat beleaguered by muntjacs from nearby Epping Forest. Secondly, all the growing is organic with special emphasis on permaculture, both outdoors and under cover within magnificent glasshouses let to them by Waltham Forest Council. Thirdly, it engages with the local population (and not so local, because people from far away make an effort to get there) in various ways including veg boxes; a local food market; a cafe; through a long list of like-minded pubs and restaurants; and by running courses and open days with one-off lectures. It also employs 17 people part-time plus 12-15 sessional workers; and many volunteers, mostly local, put in a day or half-a-day's work for which they enjoy a good lunch and get to take home any 'grade-outs'. Fourthly, Organiclea is run as a workers' co-operative in which, as its website says, 'the activities are managed by the workers directly, without the need for separate managers, owners or bosses. Organiclea is not-for-profit: any surplus is reinvested within Organiclea, or in support of other co-ops or not-for-profit organizations with similar aims.'

Organiclea defies the *idées fixes* holding back the grand cause of enlightened agriculture. We are told that people these days aren't interested in farming or growing – yet they flock from miles around to be involved in Organiclea. We are warned that people in big cities cannot be fed except at long range from mega-estates with fleets of trucks – yet the scope for peri-urban or even urban farming is obviously enormous. We are told that small areas cannot be significant employers – yet Organiclea provides about four jobs per hectare (that's including enterprise and outreach work). We are also told that the world in general needs a hierarchy of managers, but workers' co-operatives of all kinds show that, for most purposes, this is not true.

CHAPTER TEN

# Agroforestry

The word seems to have got around that trees and farming don't mix. Trees take up too much water and nutrient, says the conventional wisdom. They shade out the crops. They harbour pests - sometimes serving as secondary hosts, for example for aphids. Properly placed and spaced they can make good windbreaks - but with the wrong spacing and orientation they may create wind-tunnels. They get in the way of machinery. Combine harvesters shouldn't have to make detours around ancient oaks. Time is money. Not surprisingly, one of the first acts of pioneer farmers in temperate lands is to clear the woodland - whether it's the open kind (with big spaces between the trees) or closed (where the canopies of the trees meet overhead). How can anyone farm with trees in the way?

Foresters have been similarly standoffish. How can we grow straight, upward-reaching timber trees unless we put them close together in rows so that they don't lose apical dominance. Hormones from the top of the tree suppress the growth of side branches. If the side branches are allowed to flourish, we tend to get wood with knotty planks and tangled grain; good for veneers perhaps, but not for construction.

Overall, farmers and foresters have commonly seen themselves as separate domains. But conventional wisdom is wide of the mark. Overall, trees make life on land tolerable and abundant. They regulate the climate both on the micro and on the grandest scale. They create soil and bring nutrients from the depths. They are the principal habitat of the myriads of creatures that *are* the biosphere, and make it possible to contemplate farming of a kind that isn't just an exercise in mining and desertification. Integrate farming and trees in creative ways and we have agroforestry. Martin Wolfe of Wakelyns in Suffolk (of whom more later) says that *all* farming should be an exercise in agroforestry. Once we start to ponder the roles of trees in the biosphere as a whole, and on the farms that already practise agroforestry, we may wonder how anyone could ever have thought otherwise. In Britain, however, Defra's conspicuous failure to support agroforestry is one more reason why people who give a damn must take control.

Very possibly the world's first farms were in open woodland. The Chinese may have integrated trees into their farming at least 6,000 years ago. All around the world we still see traditional systems of many kinds that depend absolutely on trees or woody shrubs, and are built around them. Mercifully, agroforestry is recognized in some official circles: ICRAF, the International Centre for Research in Agroforestry based at Nairobi, now called the World Agroforestry Centre, was founded in 1978. In the tropics its importance is obvious (although the growing interest in agroforesty has not curbed the devastation of Brazil's dry forest, the Cerrado, to make way for soya and sugar); and as the floods and droughts in denuded Britain have been showing us of late, agroforestry is just as vital here. Once again, the affluent West needs to learn from the beleaguered South.

The trees themselves can be virtually of any kind, both broadleaved and coniferous, depending on where they are being grown and what they are for: timber, stakes, fruit, nuts, biofuel, browse, shelter, wildlife, or to bring water and nutrients up from the depths, or to purify water on its inexorable passage to the rivers, or just to stabilize the land. Ecologically it must be good to incorporate local, native trees which the native animals are adapted to; but often, in practice, exotics are preferred and sometimes they blend in well enough. New Zealand, for example, strongly favours Monterey pine (*Pinus radiata*) in agroforestry. It grows quickly and has versatile timber and is very widely grown in pure forestry plantations worldwide, although it's endangered in its native California. Many countries favour eucalyptus species for all kinds of purposes, and they of course are indigenous to Australia. Eucalypts can grow extremely fast - to a useful size in seven years - but they can also play havoc with local ecosystems. In India I have seen Australia's *Grevillea* planted as a shade tree in tea plantations. China makes extensive use of its native *Paulownia*, a relative of the foxglove in the family Scrophulariaceae, which is native to North America and sometimes known as a wonder tree. It grows up to six metres a year and yields timber that's both light and strong and used for a great many things, including sounding boards for traditional stringed instruments. Its flowers are rich in nectar and its leaves make good fodder but mature late so they don't shade crops sown beneath.

In Britain we can do very well with our native or naturalized poplars, willows, hazels, apples, pears, cherries and plums, plus oaks, hornbeams, limes and the rest. However, most of our conifers are exotics (our only true natives are the Scots pine and the yew - and perhaps the common juniper).

Trees can be arranged on farms in many different ways. Worldwide, as outlined below, farmers may simply graze their animals in natural or semi-natural open woodland. Often, trees or shrubs are planted around the edges of fields as with the English hedgerows - to mark the boundaries, enclose the livestock, and provide browse and shelter and reduce soil erosion. In the four-side system of China, belts of trees are used to demarcate and shelter entire regions. Sometimes they are planted on the tops of hills or on contours to help control the flow of water and protect the soil. Often trees - especially willow and poplar in temperate countries - are planted along river banks for wildlife habitat and to protect the banks. River planting is called 'riparian'. Sometimes trees are planted in rows with crops of all kinds - arable, horticulture - or livestock raised in between, which is called alley cropping. The trees must be orientated precisely so as to minimize shading and maximize protection from the wind (two ambitions that can be at odds). Worldwide we see trees of all kinds, arranged in all kinds of ways, integrated synergistically into every kind of farming - arable, horticulture, pastoral and even aquaculture.

## Trees with arable and horticulture

Martin Wolfe is a former professor of plant pathology who now practises agroforestry on his 20-hectare (about 50 acres) farm at Wakelyns farm in Suffolk. The system is silvo-arable (silvo is from the Latin *silva*, meaning forest) and based on alley cropping. Although it aims to pay its way it has been set up largely as an exemplar, and for research, under the auspices of the Organic Research Centre (ORC). Wheat, potatoes and other crops are raised between rows of trees about 18 metres apart (the optimum distance is still being explored). The trees include hazel and willow for short-term use - stakes, wickets, biofuel, nuts; plus fruit trees and timber trees.

The fruit trees - plum, cherry, pear, apple - are wide enough apart in the rows to reduce cross-infection, but close enough to enable pollination (by insects).

'Overall we have as many fruit trees as a fair-sized orchard,' says Martin. 'But orchards are the last place to grow fruit trees. That's when you can really get disease problems!'

The timber trees include oaks and hornbeams - a fabulously hard timber that once was used to make the hubs of wagon wheels and cogs in ancient mills. The hardwoods are a long-term investment: 'My pension,' Martin says. He also runs courses at Wakelyns with and for the ORC. His farm is a marvellous place to spend time.

The wheat grows very well between the rows of trees, and the trees too are unfazed. The husbandry is organic and cultivation is minimal and shallow and does not affect the trees' roots (which can afford to lose roots near the surface); and the roots of the cereals and the trees coexist happily. Shading is minimal - and in any case, says Martin, 'Cereals in open fields can be heat-stressed, even in Britain; and then their growth is suppressed'.

The trees at Wakelyns bring up nutrients from deep down and deposit them on the surface when the leaves fall in autumn. They also bring up water, and by their transpiration create a more agreeable microclimate, cooler and moist. Plants appreciate this as much as human beings do. Wakelyns did not suffer significantly from the droughts which so beleaguered the surrounding, conventional, open fields of Suffolk in 2012. Inputs are minimal yet yields measured over 10 years match those of the surrounding high-input farms. The cereals can also be more profitable since inputs are so low - and the timber and fruits are a bonus.

We should ask: does the presence of the trees depress the yield of crops in between? The general answer from observations worldwide seems to be no. Of course the trees do take up some space so we then need to ask, does the value of the trees make up for the loss of area? The general answer seems to be yes. In fact the real question is, do the crop and the trees together give a better output overall than either of them would if they were each grown as monocultures on the same area? Again the general answer from studies worldwide seems to be yes - whether output is measured in money, or biomass, or overall nutritional value. As noted above, the relevant measure is that of the *land equivalent ratio* or LER: the amount of land that would be required to produce as much cereal and timber, etc, if all the constituent species were grown as monocultures, divided by the actual area of the mixed system. If the LER is greater than one then that represents a *yield advantage*. In many studies of agroforestry worldwide, the LER is greater than one. The world would surely seem *bigger* if all farming was agroforestry.

Finally, as a very worthwhile bonus, the rows of trees serve as beetle banks, harbouring predators of all kinds, though mainly invertebrates, that sally into the crop and gobble up the insect parasites. Bank voles make their burrows along the edges of the rows of trees - which, ecologically speaking, are forest edge. Bumble bees nest in the burrows when the voles vacate - and they are the chief pollinators of the clovers which replace the wheat in rotations. Barn owls have returned to feed on the voles. Thus the biologically impoverished cereal monoculture re-emerges as a bona fide ecosystem.

Horticulture too can, and surely should, incorporate trees in all kinds of ways. One my favourite market gardens/smallholdings (it includes some pigs and chickens) is Worton Organic Gardens at Cassington near Oxford, run by David and Aneke Blake (see Chapter 11) - which includes a great many trees including walnuts and plums and an array of traditional, rare local apples. David doesn't think of himself primarily as an agroforester, but he is. He also grows some of the best vegetables you will find anywhere, and Aneka (who is Dutch) grows a stunning array of flowers.

## Trees with livestock

Silvo-pastoral systems - trees with livestock - take many forms. At appropriate times of year, farmers in Argentina bring their sheep and goats to graze among the great variety of trees in the open woods of the hillsides.

Closer to home, the Iberian peninsula has its forests of evergreen oaks, called *dehesa* in Spain and *montado* in Portugal. Most of the trees are holm oak or cork oak, great sources of acorns, and the forests are home to the famous black pigs that provide some of the world's finest ham. In Andalucia there are entire streets of shops selling *jamón*. The system is suffering now as cork for wine bottles is increasingly replaced by plastic and screw tops, and because of disease problems with the pigs. But to lose these forests would be tragic. Similar at least in spirit are the herds of pigs in Britain's Forest of Dean, and in continental Europe - where the pigs often take the form of wild (or fairly wild) boar.

Contrary to common lore, well-placed trees can *improve* the growth of grass and extend the grazing season by improving drainage when it's wet and providing a cooler, moister atmosphere when it's hot and dry. In Britain shelter from trees and shrubs can reduce loss of lambs by 30 per cent and in Costa Rica cattle grazed in the relative cool and shade of woodland may give up to 30 per cent more milk. As cited earlier in this book, in the Philippines Bob Orskov encouraged owners of coconut plantations to allow local farmers to introduce cattle to graze the vegetation in between the trees - and the trees yielded more heavily as a result. Chickens are natural woodland creatures too: the FAI (Food Animal Initiative) at Wytham in Oxford is now pioneering woodland poultry. English fields were traditionally surrounded by hedges - the system known as *bocage* (from the medieval Latin *boscus*, meaning wood). They enclosed the livestock and provided shelter and some browse, including ash and holly.

They were also havens for wildlife, and provided corridors between woods and a source of berries and nuts and other benison.

There are some surprising bonuses, too. Walnut trees can reduce attacks by flies; holly helps to suppress ringworm, and the leaves can be good sources of trace minerals including selenium and copper which help to boost the immune system. Trees (leaves and bark) are a rich source of tannins which help to suppress parasites. They also help to clear up surplus water by soaking it up and transpiring it away. By helping drainage, trees help to militate against liver fluke.

All in all, farm animals seem healthier and happier among trees - which is hardly surprising, since most of them, especially pigs, chickens and cattle, originated as woodland animals. Good welfare and high production don't always go together - but they do here.

## The forest garden

Then there is the forest garden - where the nuts and fruits of the trees and shrubs themselves are the primary crop. The native people of North America, or many of them at least, were farmers and gardeners as much as they were the hunters and warriors of mythology. In the eastern US a principal staple for many tribes was the American chestnut. Once it was North America's commonest broadleaf - it was said that a squirrel could hop from chestnut to chestnut without touching the ground all the way from Georgia to Maine. It was driven to extinction in the early twentieth century when Europeans tried to cross it with Asian chestnuts to increase the size of the nuts, and so imported an Asian fungus which the American trees could not cope with. The same kind of mistake has been repeated many times, and still is. It's called hubris. The native North Americans also harvested maple syrup and honey from trees and so the modern Americans and Canadians still do. All this is forest gardening of a kind.

In Britain Martin Crawford produces all manner of fruit from the trees, shrubs and understorey (the plants that grow beneath the forest canopy) on his two-acre site at Dartington in Devon - illustrating once more what can be achieved even on very small areas. Only in a seriously perverted economy does anyone need 1,000 hectares to make a living.

I would not expect or encourage anybody to fill their holding with trees on the strength of this one short chapter. But as outlined in the Resources section, there is a growing literature on agroforestry in its myriad forms, and quite a few formal courses, and almost all the farmers I have ever met have been happy to share their ideas. All existing and would-be farmers and growers the world over should take agroforestry seriously. I am inclined to think that humanity as a whole should remember its forest ancestry (albeit a prehuman ancestry) and base all our future civilizations on trees - in line with the forthcoming age of biology. But that's another story.

PART THREE

# The Agrarian Renaissance

CHAPTER ELEVEN

# The absolute importance of food culture

Those who have read this far and have experience of small-scale farming may now be asking 'But is this really possible?' For many have worked themselves into the ground, and still gone bust; and others hang on by a thread, living on debt and steadily working their way through their capital.

In truth all farming is hard, as all worthwhile pursuits tend to be, but it really does seem impossible when all the cards are stacked against it; as in Britain in particular right now, they very definitely are. The price of oil still favours big machinery (the price is adjusted to ensure that this is so) and industrial chemistry still seems to offer an easy option (though there is growing disillusionment) and modern British governments are very much aligned to the corporates (Britain's food strategy, according to one leading commentator, is to 'Leave it to Tesco'). In Britain (though not necessarily elsewhere in Europe), big landowners receive enormous handouts from the EU (a fixed amount per hectare), while those with less than five hectares are not considered to be commercial and remain hugely under supported.

Perhaps worst of all is that not enough of us give a damn. We don't put enough pressure on governments to take food and farming seriously. We allow ourselves to be conned into the belief that big-time farming and industrialization are necessary, and that they bring us cheaper food, which serious analysis shows is not the case (see Chapter 3). The British and the Americans have developed a cheap-food mentality: even those who can afford the best are inclined to opt for the cheapest, although these same people may have expensive cars and houses.

In the next chapter I will describe a plausible route whereby a great many of us - any of us in principle who are fit enough - could become farmers; the six-step path from which this book takes its title. As discussed in Chapter 3, no more than half the people should work on the land in any country, and in the most industrialized countries, probably no more than 10 per cent. So although Britain now needs a million more farmers urgently, most of us will continue to be otherwise employed. Yet, emphatically, that does not mean that the non-farming majority - perhaps 90 per cent of the population - have nothing to contribute to the agrarian renaissance. On the contrary. It's our job to create the conditions that will make enlightened agriculture possible, in which enlightened farmers can flourish. We need to recreate the right marketing structures, the right economic framework, even if our elected government is pulling us in a different direction, and the right mindset. We need to recreate a true food culture.

## What does food culture entail?

Most obviously, for starters, we need to give a damn about food: what's in it; what it tastes like; *and how it is produced.* Provenance matters. I was once assured by a food technologist that it is irrational to care how food is grown. All that matters is the chemistry, he said: whether a given food contains the prescribed proportion of protein, or vitamin B12 or whatever, and does not exceed the acceptable levels of toxicity. Damage to the biosphere would be taken care of by the market: if the damage is too costly, the practice would be adjusted. Cruelty to animals was an illusion. It's anthropomorphic nonsense to suppose that a chicken suffers and in any case, in a neoDarwininan world, it's us or them. And so on. Such arguments are morally and aesthetically vile - no basis for a civilized society.

Again, our attitude to food (which matters) seems very much at odds with our attitude to other material goods (which on the whole matter far less). One hears of people who care very much whether their bags are bona fide Gucci or Versace complaining about the price of fresh milk and buying the homogenized, dried and rehydrated milk from anonymous cows that are bred to lactate to within an inch of their lives, just because it costs 25p a litre. I do not presume to criticize. Merely to point out that this is odd.

When food culture still exists - as it does in parts of Italy and France, Germany and Austria, Turkey, India and Southeast Asia - the whole atmosphere is different: rows of little shops and cafes and market stalls, no two the same; teeming markets piled high with excellence. In such societies only the farmers (and cooks) who produce true quality *can* survive; this is where Adam Smith's invisible hand really can be felt. People at large in such societies are prepared to spend 30 per cent or more of their income on food because their whole lives are built around it, including and especially family life, and their lives are much better as a result. In Britain people may spend only 10-12 per cent per head of their net income on food and call that progress, but since many are obliged to spend 50 per cent on the roof over their heads, this works out at 20-25% of disposable income: and after we have stirred in all the taxes we must pay including VAT, food actually takes up 30 per cent of our net income. So we get far less for our money than those who try to get their priorities straight in the first place. Instead of spending our money on things that matter most, we give it to bankers.

The first and most obvious contribution the non-farmer can make is to be a good consumer: know what good stuff is, and be prepared to pay for it. Of course, many people in the modern economy can't afford to do this; but that has to do with social justice, and it's incumbent on all of us to fight for social justice. We certainly won't achieve justice by supporting corporates and industrial farming while allowing good farmers and retailers to languish, and allowing governments to shirk their social responsibilities. Alas, too many consumer movements have an us-or-them mentality. The consumers gang up to get the best price *at the expense* of the producers. Yet as discussed in Chapter 4, societies cannot function properly except by cooperation. Consumers and producers should be working together to the common goal of good food for all, in a flourishing biosphere. There are many examples of this worldwide as we will see in the next two chapters (community-supported agriculture is one good model).

The most discerning consumers are generally cooks; they know good food by working with it, sniffing it in all stages of preparation. Like farming, cooking can easily absorb a lifetime, and is well worth it. As outlined in Chapter 2, traditional cooking is the key; and when trade is fair and otherwise sensible (Chapter 3) everyone can try everyone else's traditions. Cooking is the great liberator. Cooks with access to a flame or a hot plate (microwaves are no substitute) can eat very well for remarkably little. Everyone should learn to cook. Schools should teach cooking: the ideological decision in Britain in the 1970s to stop domestic science in Britain's schools was yet another piece of state-sponsored vandalism (or misplaced do-goodery; whichever way you care to look at it). People who want to get seriously involved in the agrarian renaissance, but don't want to farm or aren't able to, should teach cooking. Carlo Petrini's Slow Food Movement, which he launched in Tuscany in the 1980s, is of huge importance, though it has never taken off as well as it should in Britain.

All farming needs marketing. We need to create new food chains, shorter and more direct than the present labyrinth. For the industrial farmer, the supermarket is appropriate: the two have co-evolved. The farmers get only a small proportion of the final price but they make up for this by producing a lot - part of the reason why the government-corporate oligarchy is still urging greater and greater production. But those who give a damn need to re-establish the marketing network that is geared to mixed local farms. Farmers' markets offer one route. Traditional cornershops and village shops provide the formal route. Small supermarkets run as farmers' co-operatives, or better still as farmer-consumer co-operatives, go one step further. If we need a million new farmers we also need a commensurate number of distributors and small-scale processors - local bakers; cake makers; micro-brewers and vintners; small butchers and *charcutiers*. It is always hard for individuals to establish new businesses but some clearly do manage it and what individuals cannot do, communities often can. Some of the ways and means are discussed in Chapter 13.

Finally, as already mentioned, food culture is enhanced hand over fist when people start growing their own, just for the hell of it. A pot of herbs on the windowsill is a step on the way. Entire families can be virtually self-sufficient in vegetables from a patio laden with containers - especially if they use the walls of the house to make a vertical garden. I seem to know more and more teachers, particularly primary school teachers, who are starting gardening clubs. Some children take to it like ducks to water. Some who hate the classroom and are banished to the corridor or otherwise excluded suddenly find their metier. Together with cookery - and, I would say, forays into the natural world - nothing can be more important than growing, and teaching how to grow, if we really care about conviviality and democracy and the state of the Earth.

Amateurs who take on an allotment, or a part thereof, soon find themselves with surpluses. With growing expertise and confidence, some may well feel that they would like to turn professional; and some of those may choose to take on livestock and more land and develop, step by step, into full-blown farmers. How this may come about is discussed in the next chapter.

## Worton Organic Garden

At Worton Organic Garden, a few miles north of Oxford, one step at a time but in less than a decade, David and Anneke Blake have created a haven on seven rented acres (a little under three hectares): on open plots and in a huge greenhouse and in polytunnels. David has a penchant for local varieties of apples and pears and heritage tomatoes, and imports vegetable seeds from all over the world. Anneke grows flowers in great variety and profusion. The pair have also established a small bakery, building their own wood-fired oven, and keep enough hens to supply customers with eggs. Recently they have added a few pigs. They sell their produce on their on-site farm shop, and some to a like-minded restaurant in town, supplementing the in-house supply where necessary with fruit and vegetables from a London-based organic wholesaler. *L'arte di mangiar bene* – 'the art of good eating' – is the slogan over the shop door.

But there is still more. From the shop grew the restaurant: its menu based on Worton produce with meat from local organic farmers and the catch of the day fish brought up from an in shore fishery at Selsey, Sussex, once a month. More still: David hosts concerts through the summer. So although people visit Worton mainly to shop, it is also the core of a true cafe society. As a bonus, the gardens, organic and diverse, teem with wildlife. On an area that mainstream agriculturalists would consider too small to register, David and Anneke employ six people.

Worton, then, is a fine example of a privately-owned small business that brings benefit to all: very much the acceptable face of free enterprise. There are comparable set-ups elsewhere and there seems no good reason why they should not become the norm, to be found on the outskirts or indeed in the middle of every town and village.

CHAPTER TWELVE

# Six steps back to the land

Britain could do with a million more farmers, at least for starters, while the US needs several million, and the countries of mainland Europe would do well at least to hang on to the ones they still have. Obviously, most of the newcomers must come from outside - people who are currently driving taxis or checking income tax or working in call centres, if they have a job at all. But there is little or no encouragement from our elected leaders and their corporate partners, indeed the opposite. For despite the odd apprentice scheme the pressure remains to cut the agricultural workforce still further in the interests of cash efficiency, however short-term. Land has become a commodity like everything else to be sold to the highest bidder, and now in Britain it is likely to cost around £25,000 a hectare. The notion has got around too - it is now a virtual dogma - that farms cannot make money unless they are huge, and the whole economy is tilted to conform to that idea. Banking and the market structure are not geared to small enterprises. So it seems impossible to break into farming at all unless you already have a fortune, and even if you did get in you are in for a rough ride, and many fail for a score of different reasons, though mostly financial or bureaucratic.

But it can be done and must be done. Despite the many difficulties, many have already set up as farmers in recent years and a lot of them, more and more of whom I am getting to know, are having a whale of time along the way - achieving the fulfilment that we all crave, and in the modern world seems so elusive.

This chapter is a lightning sketch of a plausible, six-step route into agriculture which others have trodden or are treading, and is open in principle to everyone who is fit enough; and the next chapter briefly describes some of the financial and legal mechanisms that already exist to help newcomers on their way, some of which my wife and I and friends and associates have helped to set up (through our outfit called FEA, Funding Enlightened Agriculture), outlined in the Epilogue. I do not presume to offer a formula or a life plan; merely to suggest that despite the mountainous obstacles, those who want to become part of the new generation of farmers can still do so. Nothing can be more important and nothing, for those with the aptitude, can be more worthwhile.

## Six vital steps from the city to the farm

The six-step route should not be seen as a one-way ascent to an ideal state like the pilgrim in *Pilgrim's Progress*. Anyone can stop at any stage. Neither is my suggested route the only one. I know several farmers who made their fortunes elsewhere then started their new careers with sheep or pigs or whatever because that is what they wanted to do. But the step-by-step route outlined here is possible for those with no personal fortune. Each new step is a big commitment but none requires a headlong leap in the dark. Many have trodden this path (albeit with endless variations). So:

### *Step 1: From expertise to market gardener*

No one should contemplate a market garden unless they are already good at growing (obviously!); and it is possible to become good these days if you have a garden and/or an allotment, and by reading and watching telly, through endless conversations, by watching other people (osmosis is one of the best ways to learn), and through any or many of the host of short- and long-term courses now on offer. Another much-frequented route these days is by volunteering, preferably in a wide variety of venues.

But it's a huge leap nonetheless from amateur-expert to pro. Amateurs by definition don't need to earn a living from what they do. Any money that comes their way is a bonus. But professionals must earn enough to stay in business, which means they must deliver good quantities of produce, on time and to the right standard, week after week, if not for the whole year then for most of it, in a northern climate where at best only half the year is really good for raising crops outdoors. There is little respite. Many fail. Yet many do succeed. In my experience, those who come through bring two extra ingredients which have little or nothing to do with the greenness of the grower's fingers. To whit:

- No one can seriously hope to succeed without with a sound business plan. The would-be entrepreneur must be able to envisage the cash flow in the early years when there are more outs than ins. If you have trouble with such matters, FEA should be able to help out.
- Secondly, those who succeed first ensure that they have a market - as illustrated by Ed Hamer and his business partner at Chagfood on Dartmoor (Chapter 9).

### *Step 2: Add livestock - the smallholding*

Market gardens by definition grow plants - but the *biological* efficiency of all-plant systems is *always* enhanced by the judicious introduction of animals. The animals most readily added to a market garden are the small-to-medium-sized omnivores, which mainly means poultry and pigs, although goats enjoy an eclectic diet and they, too, are good additions, valued for their hair (which, after all, includes Angora), their milk and/or their meat, which more and more people appreciate; goats have always been a prime meat animal in Asia and Africa (which more and more Europeans are cottoning on to). Add small livestock to a market garden and we have a bona fide smallholding.

The transition from allotmenteer to market gardener needs a conceptual leap into the vast and intricate world of finance in general, and marketing in particular. The decision to introduce animals raises the ante again - and adds a moral dimension. For it's a pity if plants die and crops fail, and deeply regrettable if the failure is the grower's fault, but it's a sin to allow animals to suffer and die through bad husbandry. Smallholding is commitment writ large. But for those who are so committed, and have the talent, the rewards can be excellent - social and intellectual as well as financial; as demonstrated not least by David and Anneke Blake at Worton Gardens (Chapter 11).

The agrarian renaissance should be largely or mainly built around a network of smallholdings. Smallholding should be seen everywhere as a standard career choice - as it became in Cuba after the Russians pulled the plugs on their oil. Some Cubans gave up high-flying careers to grow vegetables, and say they have never been happier. The job is agreeable and in a country that appreciates food, growers are respected. But smallholdings, alas, have largely been sidelined in this neoliberal age of high-tech monoculture.

Yet there are further steps to be taken.

### *Step 3: Add grazing*

Add grazing and we step up from omnivorous livestock to ruminants and shift from smallholding to the bona fide mixed farm. This may seem a huge transition, since we have been conditioned to think of farms as vast estates - or vast at least compared to the modest acres of the market garden or the smallholding; and the quantum jump, among other things, brings us hard up against the issue of land prices. Even at £25,000 a hectare you can buy enough land for a smallholding for the price of a fancy car. But when we get into wide expanses of grazing, the price of land looms ominous indeed.

But all is not lost. As discussed in Chapter 13, land ownership is not the be-all and end-all and it can be far cheaper, and more feasible, and for immediate purposes more sensible, to rent land. What really counts are right of use - usufruct: the right to use the fruits of the land - and secure tenancy. Many more farmers and landowners who are not prepared to sell any of their land are willing to rent out parts of it, and rents are cheap compared to the capital cost of land.

Good things can be achieved even on remarkably small acreages, and so there is a good case for restoring the house cow, to provide at least enough milk for an extended family or a club - which can be forthcoming all year round if there are two cows, one calving in spring and one in autumn. The micro-dairy or the once ubiquitous small dairy could and should again become the norm, as discussed in Chapter 8. At North Aston in Oxfordshire, Matt Dale made a reasonable living from 17 Ayrshires, selling direct; and as described in Chapter 8, Nick Snelgar has now set up a micro-dairy in Hampshire. Nick learnt a lot from Matt but his business model is somewhat different. No two are ever quite the same.

Nick aims to have around 20 cows but Matt has now concluded that 40 is probably the ideal number which, on good grazing, would need a minimum of 80 acres. 80 acres would cost up to £800,000. This is a lot, but is comparable to the price of a fairly modest house in south London, and a somewhat less modest house in Oxford. Now could be a good time to rent land, for a great many farmers in Britain are on the point of retirement and are happy to rent out such acreages. Neither is dairy farming the only option. We have friends and acquaintances who keep small flocks of sheep for meat and wool, or else raise a few bought-in calves to slaughter weight each year on five acres or less; not as whole living, but as a contribution to the family budget and, if enough people did it, to the nation's food supply. In much of the world, not yet submerged by neoliberal corporates, such small-scale production by part-timers is the norm. If several people team up to run a farm like a small company, and some of them keep their regular jobs, then, even in finance-ridden Britain, small-scale farming becomes a realistic option.

Ideally, grazing with a small flock or herd of sheep or cattle (or geese) is properly integrated with the smallholding. In medieval times the two were generally separate and the peasants brought in dung from the fields to put on the vegetables. Nowadays with the aid of electric fences (appropriate high tech again, in the service of craft farming!) it is possible at least in theory to allow the ruminants to graze the green manure crops that break up the rotations. With cunning forms of agroforestry including alley cropping the whole thing can become geometrically rather neat. Ideally, I reckon, horticulture with poultry and pigs would move *en bloc* around farms that primarily were arable or pastoral, the crops benefiting from the livestock that were there before, and the livestock making life harder for weeds and breaking cycles of disease, and so on. Then, truly, we have synergy.

Alternatively (or in addition):

### *Step 4: Add arable*

Again, as things are, with wheat, barley, maize and rape grown in monocultures as far as the eye can see, the leap from market garden or smallholding to arable seems too far. But as suggested in Chapter 6, judicious small-scale arable can make very good sense (especially when linked to a local bakery or a micro-brewery); and again, as with livestock, it is easy to see how arable can be integrated into a horticultural rotation. If the arable patch is the same size as the market garden, then the two could swap positions wholesale every few years, which would be good for both. If livestock were included in the rotation we then have at least one version of the fully integrated mixed farm: the agroecological ideal.

### *Step 5: The full-blown agroecological mixed farm*

This is what we should be aiming for: cereals, pulses and tubers; fruit and vegetables; livestock of several species and all classes; trees; ponds; all integrated with each other and with the wild surroundings to make a synergistic whole, a true ecosystem: resilient, sustainable, productive, humane. Horton Grounds Farm in North Oxfordshire, owned and run by Catherine and Graham Vint and their business partner Simon Thomas, provides a fine example (Chapter 8).

Farms of this kind seem more or less to have arrived. They demonstrate what enlightened agriculture can mean within a temperate country. Yet there is one more step that can and, for all kinds of reasons - logistic, financial, social - *should* be taken. So:

### *Step 6: Farms as communities*

The full-blown, agroecological mixed farm, as described in Step 5, has one or two obvious drawbacks. For the traditional small mixed farm is based on the idea of the single owner or tenant: the yeoman or peasant farmer and his family (in traditional Britain it was usually 'his'). But the lone farmer on a complex farm must have many different areas of expertise and must work extremely hard to keep all the balls in the air. As history shows, lone farmers, however good they are, have been extremely vulnerable: economically, politically, and indeed physically, sometimes suffering catastrophically from flood, drought, cold or whatever. So, while retaining the principles of polyculture and low input, the structure must be made more robust.

As always, the way to achieve this is to involve more people, working in concert: sharing liability and risk; supplying complementary skills; probably operating quasi-independently day-by-day but sharing and cooperating whenever this is appropriate; a network or an ecosystem of interlocking enterprises working as one unit. Socially the complex farm can be conceived as a hamlet, a community with common interests and a common goal. The overall holding, with several or many people on board, may no longer be small, at least in area. But each enterprise within it remains human-sized, with plenty of tender-loving care. The farm-as-hamlet model differs from the commune because each player within it retains his or her autonomy; not a merger of effort and personality but a co-operative of the free: James and Henrietta Odgers' Stream Farm in Somerset (see page 188) illustrates the principle beautifully.

Communes can be good too, but they involve extra layers of commitment and often of ideology and require their own discussion. Many have worked wonderfully and many have fallen apart. The hamlet model should be self-renewing, ever-changing yet staying essentially the same, like a coral reef.

There are more and more examples in Britain and the world at large of farms run and owned by communities: usually by local communities (villages, neighbourhoods) but sometimes by communities defined not by geography but shared interest. Nick Snelgar also helped to establish a community farm in Martin. The local people run the farm themselves - livestock as well as horticulture; and sell the surpluses in the village shop, also run by the community. In Todmorden on the borders of Yorkshire and Lancashire the townspeople use virtually every space to grow fruit and vegetables, including the small patches of land in front of the police and fire stations. Three wonderful examples of local people working in partnership with professional farmers and growers for the common good are described in this book: Tamar Grow Local in the Tamar Valley between Devon and Cornwall (see page 190); Organiclea in the Lea Valley in Essex (Chapter 9); and Growing Communities, a combination of allotments run by members, and farmers and growers in north London (Chapter 13). The benefits of all these enterprises, including social benefits, are enormous.

Carlo Petrini, founder of the Slow Food Movement, conceived the notion of the convivium: people just getting together and starting something new. The 'something new' could be a small farm, owned by groups of friends and run by them, either full-time or as part-timers. Ideally, the farmers in such a set-up function as a team, like the staff of a magazine, or the players in a football club: bound by a professional ambition, but creating a sense of family. In such a vein, Sam and Lucy Henderson and friends set up Whippletree Farm in Devon, combining smallholding with grazing.

Those who would bring about the agrarian renaissance that the world so badly needs must think flexibly. We need to break down the boundaries between producer and consumer, or professional and amateur - as exemplified by Organiclea and Growing Communities. We need to acknowledge the importance of part-timers, who combine farming with other jobs - anything from fisherman to builder to doctor to academic to accountant to journalist or what you will. Indeed, part-timers of many different kinds already produce a huge proportion of the world's food. Many regard it as the ideal way of life: each strand of career providing a break from the others.

We need, too, to break away from the yeoman farmer model. At least, yeoman farmers certainly have their place, but in today's world there are many more possibilities. We ought to think of farms not as personal fiefdoms but as joint endeavours. It is often said that people won't put their best efforts into farms unless they own them, and feel that they can pass on the fruits of their labours to their children. But this, like so much accepted wisdom, just isn't true. Certainly some of the best and most dedicated farmers I know are not owners but managers. Their satisfaction comes, just as it does in all worthwhile professions and trades, not from creating a dynasty but from doing the job well, and being acknowledged by their peers. School teachers and general practitioners don't have to own their places of work to be remembered and revered.

But whatever route the farming newcomer goes down, he or she will have to deal with the sober practicalities of life: legal and financial. We will look briefly at what this entails in the next chapter.

## Stream Farm, Broomfield, Somerset

James and Henrietta Odgers' Stream Farm in Somerset is conceived not simply in the owner-farmer yeoman tradition or as simple tenancy but as a 'share farm'. Several enterprises operate on the same piece of land independently but also, when it matters, co-operatively. The farm thus becomes a convivial, synergistic community, with benefits all round. A share farm can be highly productive, in food and money, while creating a lot of satisfying jobs – true careers – and so regenerating rural economies.

As is so often the case, James came into farming after a successful career elsewhere. He bought Stream Farm – 250 acres of the Quantock Hills – 13 years ago. He began with a suckler herd of 25 Dexter cows (Britain's smallest breed – the size of a Shetland pony) and 27 Hampshire Down sheep. Now the farm is organic, and mixed. It has 70-80 suckler cows and 250 breeding ewes, plus pigs (mostly crosses of Gloucester Old Spot, Berkshire, or Saddleback for pork and sausages), and the farm also buys in one-day-old Devonshire Gold chicks and sells 125 chickens a week. There is an orchard of 1.5 hectares (around 3.5 acres) with five varieties of apples that are sold as juice; plus hives for honey; and three ponds, aerated by a stream, for raising rainbow trout – many of which are cured in the on-farm smokery. For good measure, the farm sells its own spring water.

Each of these enterprises is run as a quasi-independent business by its own farmer – although, at busy times, they all cooperate. The farmers put in the labour, while James supplies the land and bears the capital costs. This is what makes it a share farm. James helps with the marketing, with all the produce sold under a common brand. James then takes a proportion of the gross profit. It takes time to build a business as complex as this and in 2014 James reckoned that the model was just about up and running. He still has to buy in some inputs but estimates that the farm is now 90 per cent self-sustaining.

James is a committed Christian and for him Stream Farm is very much a moral and social exercise as well as a commercial one. He aims both to see the farm run to the highest standards of agroecology and to help bring newcomers into farming. His goal is that Stream Farm should accommodate 10 quasi-independent businesses at any one time – one for every 25 acres, although several of the businesses rotate on the land. Each should earn up to £20-£23,000 gross per annum – and generally, after a few years, James encourages them to move on and set up on their own. Stream Farm thus functions as an 'incubator farm'. Ideally the newly set up enterprises then become part of a growing franchise, all selling their produce under the Stream Farm label.

## Tamar Grow Local

Tamar Grow Local, founded in 2007, is very much in the spirit of enlightened agriculture: producing a great variety of food for local sale and consumption; involving the whole local community – including professional farmers and growers, and also the hobbyist producers of vegetables and fruit and honey; providing training (in beekeeping, fruit grafting, pruning orchards); and seeking always to conserve the unique landscape of the Tamar Valley.

The river Tamar itself runs north-south, dividing Cornwall to the west from Devon to the east, and entering Plymouth Sound to the west of Plymouth.

For more than 200 years the valley banks were a prime source of garden produce for the west country and beyond, just as the Lea Valley once supplied London. But the Common Market, as it was then, opened the gates for produce from sunlit Europe and the Tamar gardens largely faded away.

TGL sees itself as an exercise in community-supported agriculture and is set up formally as a community interest company (see Chapter 13). Growing enterprises for non-professionals include community allotments, TGL's first-ever project, plus community orchards. Wherever possible the professional growers are linked into co-operatives – with everything from livestock to fruit juice and honey. They share equipment and transport, and run stalls on a rota at farmers' markets. TGL includes co-operatives for bulk seed purchasing and wood fuel.

Food distribution is being brought to a fine art. Producers don't have to take their goods directly to the market but deliver it (sharing transport wherever possible) to a series of food hubs where individual customers and small retailers can pick it up – many having first ordered online. Some of the hubs link to railway stations so commuters can pick up their goods on the way home. Once a month the growers make use of the river Tamar itself, loading their goods on to an old Yealm crabber Shamrock as they did in the mid-nineteenth century, and so to the Royal William Yard market in Plymouth. There is also a home delivery service: £2.50 per delivery into Plymouth; £1.50 for the rest of the Tamar Valley.

Organizations of such complexity cannot be installed overnight or imposed as a package. They must be brought into being by a kind of guided evolution – a clear vision of what's needed, geared to the local realities. From the outset TGL has been steered and managed very ably by Simon Platten, aided by volunteers and now by three full-time staff made possible by various grants. TGL shows what communities can do by working together when they find the right leaders. We don't have to let the corporates run our lives.

CHAPTER THIRTEEN

# Land, money and the law

To create the kind of world we need, and most of us surely want, we must be radical. The present way of farming with its underlying precepts, economic, moral and political, driven by competition for short-term wealth, just will not do. Specifically we need small-to-medium-sized, diverse, low-input farms and corresponding markets, the very opposite of the prevailing model. More generally we need to be aiming not for bigger piles of wealth in fewer hands but for general conviviality and a flourishing biosphere, and to achieve this we need to liberate the genius of the people: experts and intellectuals on tap but not on top, as Winston Churchill put the matter, with politicians who truly see themselves as public servants. All this requires radical change to the zeitgeist.

Yet it's a mistake to be more radical than we need to be. That wastes time and energy and leads to unnecessary strife. Notably, we do not need, as was commonly supposed in the 1970s, to 'smash capitalism'. Capitalism is an elusive beast which takes many forms. Some of them are repellent even to traditional US Republicans or British Tories, though both are perceived as the natural parties of capitalism. As the Tory Prime Minister Edward Heath said in the early 1970s, capitalism in some guises is 'unacceptable'. Neoliberalism, which did not come properly on board until after the 1970s, has proved to be the most unacceptable of all.

But *some* manifestations of capitalism can be benign. Many of the mechanisms that are associated with capitalism, properly deployed, can serve society very well, and may be just what we need. These mechanisms are able, at least in theory, to strike the essential balance between the needs and reasonable aspirations of free individuals, and the needs of the society and the biosphere as a whole. What, in practice, does this entail?

## The benign face of capitalism

Over the past few years three, simple ideas have come to the fore which between them seem able to ensure that finance can indeed work in the best interests of us all, and not simply for the few:

### *1. Social enterprise*

Social enterprises do as the name suggests. They are true enterprises, intended at least to make a modest profit and certainly not intended to make a loss: they are not conceived primarily as charities (although they may have the legal status of a charity). Their prime intention is not to make their perpetrators rich but to provide some benefit to society as a whole or to the biosphere - helping, for example, to finance a school or a wildlife reserve. This is commerce in the service of values that are not purely commercial. Social enterprises may take an infinity of forms. A brilliant example, OrganicLea, north-east of London, takes the form of a workers' co-operative (see Chapter 9).

### *2. Ethical investment and positive investment*

Ethical investment means that people at large can give their support to enterprises - which in general would be social enterprises - that they believe in; although Jamie Hartzell of Ethex (see later) prefers the term 'positive investment' to 'ethical investment'. He points out that ethical investment *can* have a negative connotation: not investing in enterprises you positively disapprove of, such as the arms trade. Positive investment means that (perhaps with the aid of suitable investment companies) you place your money only with companies that you positively endorse, and want to take an interest in - such as local farms of an enlightened nature.

Ethical (or positive) investment may seem an innocuous idea but there is more to it than meets the eye. The imperative in all economies is to keep money circulating. While it is circulating it can do good; but when it simply sits in coffers it contributes nothing unless, like a miser, you revel in the simple knowledge that it exists. It doesn't really matter from a purely economic - in the sense of financial - point of view what the money is spent *on*, as it buzzes around. As modern governments seem to agree, money spent on arms contributes to the perceived wealth of a nation, and hence its economic growth, just as much as it would if spent on housing. But from a social and a biological point of view, it is obviously better to spend the circulating money on good things than on bad things, and on things that don't make a mess, exhausting soils and polluting seas and driving other species to extinction.

Money spent on excellent food produced by the methods of agroecology with respect for food sovereignty, and on looking after the biosphere, serves all purposes very well. It does good as it circulates - and, if it is invested wisely, it does not pollute the world but can leave it more fertile. So money spent on excellent food and enlightened agriculture and all that go with them should be win-win, economically as well as socially - but not, paradoxically, if the agriculture is designed *expressly* to make money in the short term, as is the present condition. Then the money stays in the hands of a minority, with huge collateral damage.

With all this in mind, a group of us in Oxford in 2012 launched Funding Enlightened Agriculture, now known simply as FEA, to act as a broker to introduce potential funders to would-be enlightened businesses, and providing the latter, where necessary, with mentoring and whatever is needed to improve their business plan and become investment-ready. There is also a variety of more general ethical investment companies, such as Gaeia, based in Manchester, and the online platform Ethex, based in Oxford; and ethical banks such as Triodos, Bristol.

In the proper spirit of cooperativeness, representatives from all the companies cited above are amoung the FEA's advisers. Those who want to begin new enterprises of a social nature will find that there are a growing number of organizations and individuals able to advise on structure and funding. Ethical/positive investment is still a minority pursuit but it could and should become the norm. One tremendous advantage - in contrast to the mega-investments of insurance companies, say - is that it is fine-grained. It is able at least in theory to deal with very small sums of money so that everyone who has any spare cash might become an investor, and to be an investor is to be involved (to a greater or lesser extent, depending on the nature of the company). Jamie Hartzell, founder of Ethex, treats positive investment as a most absorbing hobby. But for many, social enterprises can offer good returns on investment.

### *3. Community-supported agriculture*

The third essential idea, more a general principle, is the absolute importance of community, as touched upon in Chapter 9. One manifestation of this is community-supported agriculture, or CSA; individual farms that enjoy community support are known as CSAs. In the US there are well over 1,000 CSAs while in Britain there are over 100 which take any one of four main forms:

- As in the US (and in France), a professional farmer offers a share of the produce in exchange for a fixed subscription that is usually paid in advance so the producer has a guaranteed income. The farmer and the subscribers share the risks. Chagfood has this kind of arrangement.
- The community itself may set up the farming enterprise, and the labour is supplied by volunteers and/or by professionals who become employees.
- The enterprise is owned by the community who form a co-operative, or something similar, and work in close partnership with existing producers to ensure a steady supply of produce for the community and a steady market for the co-operative. Cultivate Oxford is a good example of this kind of arrangement.
- Finally, communities may invest in a farming enterprise - but the enterprise does not necessarily trade primarily with the investors. The investors receive a dividend and/or enjoy special privileges on the farm and also have the satisfaction of knowing that they are taking part in something worthwhile. To ethical investors, this sense of satisfaction can be just as important as the possible financial gain.

Clearly, in CSAs all parties benefit. The farmer has a secure market - a luxury these days! - and should be paid far more per item than a supermarket would pay. The consumers benefit because they get exactly what they want, all fully traceable, and they need not pay more than they would in a supermarket because they are saved the cost of the convoluted food supply chain that snaffles 80 per cent of the supermarket retail price. CSAs, in short, are a fine example of practical community action; and show how the conventional mechanisms of finance (investment, loans, etc) can bring direct social benefit. All in all they provide a fine example of social democracy. At the very least, CSAs are a vital stepping stone to a more secure and convivial future.

Producer-community partnerships cannot work, however, unless there are enough producers; and especially enough small farmers, flexible enough to fit what they do to the community's demands. As things stand, at least in the ultra-urbanized countries that are considered advanced, we need a lot more farmers. But how can anyone who isn't already a part of a farming dynasty, get into farming? What are the first steps?

## Getting started: finance

Ideally, businesses are financed internally - by selling their produce for a profit that is at least large enough to keep going.

But businesses don't become self-financing from the word go; and they often experience setbacks even when they are up and running. All, at some point, need *external* funding as well. All of them need it at the start-up stage. Many need external finance in their infancy, after the business is installed but before it is in full flow; but money for infant businesses is often harder to come by than money for start-up. A recent, excellent, but unfortunately little-known study from Sustain called *Financing Community Food* revealed that a huge proportion of small businesses begin on a wing and a prayer, with far too little knowledge of what's entailed, or how to raise finance, or how to put together a business plan to help attract finance - ie to become investment-ready. For starters, it really is worth trawling a few websites. (I list a few in the Resources section and each should lead to others). When you do get to start-up point, it's worth finding a mentor to guide the business through the early stages. FEA can help in this; or, if it can't, it can recommend people who can.

External funding, including start-up money, comes from four main sources, and businesses of the kind we are talking about here should strike a balance between them all, depending on their needs and their stage of development. So:

### 1: *Donations*

Many small businesses begin with money given by relatives or friends. In practice there is no clear dividing line between donation and loan - 'Pay me back when you can, son!' - or between loan and investment. Michael Norton, co-founder of the crowd-funding platform Buzzbnk (see later), observes that loans often turn into donations, for human beings are not the obsessively acquisitive beings envisaged in classical economic theory. Many are happy to lend money without any serious expectation of getting it back, just for the pleasure of being involved in something worthwhile. But as the Sustain report remarks, many would-be entrepreneurs are just too shy to ask people with money for their help.

### 2: *Loans*

While donations are usually modest, loans can be for any amount. But loans don't always or usually take the form of quasi-gifts, or of gifts in disguise. Most lenders want their money back, usually within a stated period, and they generally want some interest. The usual source of loans, the high street banks, are extremely reluctant to lend to small farms these days (or, as many complain, to small businesses in general) and although this seems to abnegate their purpose, their raison d'etre, this, alas, is the way things are. Loans the banks might conceivably make must generally be secured by collateral: for example, the borrower might be expected to sell their house to pay their debts if the business fails. Ethical banks, such as Triodos, should be more helpful. But in general, loans via established, formal routes are not easy to come by these days (though they were often too easy to come by before the crash of 2008).

### *3: Grants*

The third main source of funding is obtainable in theory from all kinds of grant-giving bodies from the Lottery Fund to the Esmée Fairbairn Foundation. Grants can be for any amount and - joy of joys - the grant-giving bodies do not expect their money back. They do, however, expect results.

Grants come with stipulations including criteria of success that must be achieved by certain dates or the grant may be withdrawn. Grant-giving bodies also have their own agenda - they don't just give money to anyone for anything. So those seeking grants must first prepare documents showing why and how their particular enterprise conforms to the grant-giver's ideals, which of course is time-consuming and can be costly. Applicants nearly always have to adjust their own plans to meet the grant-giver's demands, and this can compromise the enterprise - in which case, even if a grant is offered, it may be better to turn it down. Grants, too, always have a limited time-run, and many a recipient has found that their business can't support itself when the grant runs out. Some just go on applying for more and more grants and so, if they get them, become grant-dependent - which for enterprises that aspire to be businesses, is not the idea at all. Others just fold up, leaving the grant-giving body with a sense of waste. The point of grants in this context is to enable enterprises to live without them.

### 4: *Investment*

External funds can be provided by investors who generally expect a share of the equity, where equity means the material assets of the company after the debts have been paid off. Sometimes, though by no means always, investors also have a say in how the business is run. There are many mechanisms of investment, all with their pros and cons, and all are legally defined with their codicils and conditional clauses. Would-be investors should get at least some idea of what those terms are, and anyone seeking investment should certainly do so. Type 'forms of investment' into Google to get the ball rolling. companies like Gaeia and Ethex, cited earlier, make life a great deal easier.

But start-up farmers seeking investments have a special problem. For although some food-orientated enterprises can be highly profitable, farming usually isn't, and returns on investment tend to be low. Farm enterprises offer lower returns than bullish companies on the stock market, but they also tend to pay less than many an ethical investment. Right now, for instance, renewable energy in the form of windmills or solar panels often offers returns on investment of around 6 per cent, whereas farming of the kind the world really needs rarely returns anything like this (I don't know of any that does). So investors in enlightened farming must have other motives; they must truly want to use their money to help create a better world. Fortunately, many do, and many more surely would if they were aware of the opportunities, and if those who seek investment would make themselves known. One of the prime tasks for our own outfit, FEA, is to broker meetings between would-be entrepreneurs of enlightened farming and would-be investors.

An increasingly popular way to raise donations, loans and, to some extent, investment is via **crowd-funding**. The would-be entrepreneur makes a short video to introduce his or her enterprise, as lively and enticing as possible, and then presents it on the net on a crowd-funding platform such as Buzzbnk. (Michael Norton wanted to call it Buzzbank but couldn't because it isn't a bank. So he dropped the a.) At its most simple, the supplicant specifies the amount that needs to be raised, and the time frame, and offers benefits (such as a box of vegetables, a weekend away camping) in return for money. If the amount is reached in the allotted time then the entrepreneur keeps the money and goes ahead. If not, then the money that's pledged is not claimed. Buzzbnk can offer repayable finance in the form of a loan. There are also crowdfunding platforms specialising in equity crowdfunding, Ethex being one.

## Legal structures for social enterprises

Next, a brief word on legalities. All bona fide companies must adopt some recognized legal structure, of which there are about half a dozen main kinds. Two are particularly pertinent to social enterprises:

- The **community interest company**, or CIC (pronounced 'kick'). The directors of a CIC can be paid, and private investors can receive dividends, but the overall benefit must be to the wider community, and all the assets of a CIC are securely locked to ensure that this remains the case.
- The **community benefit society (CBS)**, which came into existence in August 2014 (along with the co-operative society) with the Co-operative and Community Benefit Societies Act. A CBS is run for the benefit of the community at large. A key feature is one member one vote. Thus (depending on the rules of the society) a person may become a

member by investing as little as £1, but he or she then has the same voting powers as someone who invests £100,000. This helps to ensure that the enterprise remains true to its purpose. No one can muscle in and rule the roost just by buying up the majority of shares. A CBS is managed by a committee and officers on the members' behalf.

## Land

Then, of course, there's the matter of land: for farmers, the *sine qua non*. The subject is huge (none huger) but here are a few salients:

In many and perhaps most countries the land is owned by a remarkably small number of people - and nowhere is worse than Britain, where fewer than one per cent of the population (in fact 0.69 per cent) own about 60 per cent of all our land. Farmland in Britain is absurdly expensive - at the time of writing it's commonly around £25,000 per hectare. Today's industrial arable farms often run to 1,000 hectares, each of them tying up a cool £25 million.

So how can newcomers break in? In the long term, meaning as soon as possible, we need serious land reform: a wholesale, concerted effort to return land ownership to the community at large - but bearing in mind (see Chapter 4) that community ownership does *not* mean state ownership! We need to reopen the whole question of land ownership - beginning by asking whether anyone should own land at all. We need to remove land speculators from the scene (another failure of the so-called free market) and to bring the cost of land more into line with what it is actually used and needed for. No major political party seems to have any particular interest in land reform but any one that did take it on would deserve to be taken seriously. But even in the short term, working within the status quo, there is a great deal that can be done, and is being done.

First, we see again the absolute importance of communities. What one person alone could not contemplate, several or many together might take in their stride. Also: although the 1,000-hectare arable farm is becoming commonplace, smallholdings and small farms can be far more productive in terms of food per unit area and can provide good livings when there are suitable markets. A three-hectare holding, or even less, can provide enough fruit and vegetables, plus eggs, chickens, ducks, pork, bacon and goat's or even some cow's milk, to make a significant difference to a fair-sized village. One person alone may find it hard to buy three hectares - probably at least £80,000 if it's near to human habitation - but it's not a huge sum for a village to find, or an urban neighbourhood. The holding may be run at least in part by the villagers or townspeople themselves (as at Todmorden), or employ a professional farmer or grower to run it on the villagers' behalf, which is the common course for CSAs; or a combination of the two. If village or neighbourhood farms became the norm, as they could, they would make a huge difference to the shape and structure of Britain's entire food network.

On the larger scale, there are various movements worldwide for people at large, including some special interest groups, to buy tracts of land and put them in trust to ensure that it is used for good farming in perpetuity - never again to be a mere commodity. In France, the Terre de Liens movement has raised 36 million Euros since it began in 2003 and acquired 2,300 hectares that it has divided into 100 farms all over the country, dedicated to organic peasant farming. TDL now involves more than 12,000 people, advises more than 1,000 aspiring farmers every year and has formed tens of partnerships with local authorities. There are comparable movements in other countries including Spain, Italy, Romania and the US; while in Britain, Martin Large and others have established the Biodynamic Land Trust; and the Soil Association also has its own Land Trust. The Ecological Land Co-operative based in Lewes, East Sussex, buys tracts of land and then breaks them up into smaller units to sell to newcoming smallholders. Famously, in 1997, the inhabitants of the Isle of Eigg in the Inner Hebrides of Scotland, in partnership with the Highland Council and the Scottish Highland Trust, bought their entire island - a community buyout: excellently documented by Alastair McIntosh in *Soil and Soul*. Other communities since have followed suit. The Scottish Crofters, a very important group of farmers, sociologically, economically and ecologically, are also buying tracts specifically for crofting.

Fancifully perhaps - but plausibly! - we might at least note in passing that the British people as a whole could fairly easily afford to buy *all* of Britain's farmland - which again, to emphasize, should be under community control and not specifically under state control. After all, we have only 18 million hectares; and (just to keep the arithmetic simple) a population heading towards 72 million, so that's only a quarter of a hectare each (which doesn't sound much but is easily enough to make us self-reliant). So a complete buyout would cost about £6,500 per head at 2015 prices. Few could afford to produce £6,500 from their back pocket - but that is what banks are supposed to be for. A bank loan for that amount (or a commensurate tax) spread over a lifetime would be trivial; the current cost of two terms'-worth of university tuition, and about a twentieth of what many people now spend on a mortgage, often for a very modest house. In reality, we would not need to buy *all* of the land. Some is already in benign hands and some is very well run and our approach to takeover should be minimalist (radical but not more radical than is necessary). For £3,000 a head we could make a huge difference.

But ownership is not the be-all and end-all. It can be a hugely profitable investment but it also brings huge liabilities, legal and financial - and social and environmental too if landowners take their responsibilities seriously. More important for farmers who just want to farm, are the principles of usufruct and security of tenure. Usufruct is a feudal concept: a landowner may own the land but somebody else may in principle have the right to the timber or the grass that grows on it, or the oil or the water that lies beneath, or the fish in the rivers.

Security of tenure means just that. Ideally farmers should be sure that they can stay on the land for long enough to develop serious enterprises and see their work through to fruition. At present in Britain, sometimes because landowners want to stay flexible so they can sell their land when they want to, contracts are typically offered for just a year at a time - so the farming such as it is must be done by contractors, sweeping in to raise a crop of carrots, if the land is suitable, or by local farmers grazing their livestock for a few months. There is no chance to build the integrated farms and farming communities that Britain and the world so desperately need. Again we see the destructiveness of economies based on short-term profit.

It is sometimes argued that farmers would not be content with mere security of tenure. The British myth has it that farmers do not farm properly unless they own the land and can pass it on to their sons and daughters, building a perpetual dynasty like the feudal families of old. But this simply isn't true. Most of the farmers I know simply want the opportunity to be farmers. Some of the very best are farm managers - employees. They want to do a good job just as university professors want to do a good job even though they don't own the university and are content to be remembered for a life well lived. Of course it is vital that everyone should be compensated at the end of their careers for whatever capital they have personally invested, with a decent pension and a place to live. Too many farm workers over the years have simply been abandoned. But that is a matter of simple justice. It has nothing specifically to do with land ownership.

With such matters in mind, various thoughtful landowners have sought to invite farmers, including newcomers, to set up autonomous or semi-autonomous enterprises on their land. Advice on how to help such arrangements proceed in a structured manner is contained in the *Land Partnerships Handbook*. By such means, without a huge and dramatic political shift, the practical goal can be reached - of small mixed farms offering plentiful employment and, if all goes well, with an agreeable sense of community.

Nevertheless, we do need changes in the planning laws. Arable crops don't need looking at hour by hour but horticulture and livestock need close scrutiny and hands-on attention. Farmers who operate complex systems must live on site - or at least, life is many times easier if they do.

But it can be very difficult even for established farmers to get permission even for a building that is theoretically temporary. Planners are afraid of the thin end of wedges - ostensible barns that evolve surreptitiously into gracious homes. Land held in trust could eliminate such misuse. All this is just a lightning sketch of possibilities. The websites cited in the Resources section will lead you into more. The world as it stands, and certainly the economy of Britain, is hostile to the kind of farms and markets the world really needs. But there are appropriate legal structures and helping hands out there nonetheless.

## Growing Communities: a social enterprise in east London

Growing Communities (GC) is not a single farm. It is a movement, set up in 1993 by Julie Brown and two friends in Camden, north London, as a social enterprise and very much as an exercise in food sovereignty, to give local people control over their own food supply. GC soon relocated to the east London borough of Hackney, where it is still based.

GC buys in produce from nearby organic farms, including its own farm in Dagenham, east London, set up in 2012. GC can use its community status to negotiate a good (but fair) price, and the growers are guaranteed a good market. Thus GC in effect is a co-operative between producers and consumers. As is the case with true cooperation, this is win-win.

The produce is then distributed in two main ways. First, nearly 1,000 families (in 1993 it was 30) who are GC members now receive weekly veg. Secondly, veg and fruit are sold to the world at large in the Stoke Newington farmers' market, which attracts 1500-2000 visitors every Saturday. GC, overall, now provides 24 part-time jobs – paid for by income from the box scheme and the farmers' market.

The members of GC raise some of the produce – including the award-winning Hackney Salad – on three Soil Association-certified organic market gardens leased from Hackney Council. Volunteers and professionals together do the growing – a part-time grower, an assistant grower; and a series of apprentice growers, four at a time, who do six-month placements. Many of the apprentices go on to raise vegetables for GC on small sites, including gardens donated by local residents, churches and the council – sites which between them form a patchwork farm across Hackney.

In 2010, Growing Communities set up its Start-up Programme, offering an online toolkit and mentoring for other groups around the UK to set up their own community-led box schemes, using the GC model. To date, nine box schemes are up and running with another five due to launch, so Julie is realizing her dream: 'I love seeing organizations like Growing Communities springing up all around the country, supporting local farmers, creating jobs and providing communities with great, local, sustainable food.' And of course, reclaiming what ought to belong to all of us: control of our own destiny.

EPILOGUE

# The agrarian renaissance and the Real Farming Trust

Can people who give a damn turn things round before the world is compromised beyond redemption? It's all too easy to be pessimistic. The prevailing oligarchy of corporates and governments like Britain's, and their chosen advisers, is wreaking havoc yet seems to grow stronger - not by conspiracy but by natural selection. Those who subscribe to the status quo are rewarded and gain more influence: those who take a stand are sidelined.

But people worldwide are fighting back - enough to suggest that sea-change is possible. More and more people, it seems, perceive that agriculture of the neoliberal-industrial kind is not working. It can never provide good food for all because it isn't designed to, and will always be destructive because the biosphere is seen merely as a resource. The enlightened kind is complex but the necessary theory - based on agroecology, food sovereignty and economic democracy - is growing ever more robust; and at the same time, more and more individuals and groups are taking matters into their own hands - developing new approaches and showing that people working together can move mountains.

For my part I have been thinking and writing about food and agriculture for more than 40 years and when my wife, Ruth, started to get seriously involved about 10 years ago we (especially she) started to turn the polemic into action. Around 2005 I coined the expression 'enlightened agriculture' which is sometimes shortened for PR purposes to 'real farming', and soon after that, with help from a kind friend (see Acknowledgements), we set up the Campaign for Real Farming with a website, www.campaignfor realfarming.org, that serves both as a forum and as an archive.

In 2009 the agricultural writer and small-time sheep farmer Graham Harvey suggested that we should establish a conference to challenge the Oxford Farming Conference, which has met every January for the past 60 years to present the establishment view of food and farming and basically to tell us that all is well. So in 2010 we started the Oxford Real Farming Conference, ORFC, organized primarily by Ruth. Eighty or so people, including a fair slice of farmers, met for a day in a medieval library. January 2015 saw the sixth ORFC – with 750 delegates, about half of them farmers, gathered to take part in four parallel streams of discourse over two days in Oxford's magnificent town hall: a ten-fold increase since our opening foray. Now we are seeking to build on the ORFC with one- or two-day focused seminars and discussions on particular issues.

Then in 2012 we decided we needed to attract benign forms of finance to the various enterprises we were getting to know about – and so set up Funding Enlightened Agriculture, FEA, which brings together would-be social funders with promising new farms and related ventures. The following year, we brought all these initiatives together under the newly established charity, the Real Farming Trust, RFT.

My personal dream, now being realized, is to establish the College for Real Farming and Food Culture (CRFFC) to discuss, research and develop the necessary ideas and techniques that are needed to bring about the agrarian renaissance and put the world on a new footing; and to teach the necessary ideas and skills at all levels from primary school to post-grad. So far I have been getting the ball rolling with lectures and discussions in venues that range from universities and cathedrals to pubs, barns and village halls. In September 2013 Ruth and I organized a one-week course on Agroecology at Schumacher College in Devon. Logically, we will aim to combine the events of the CRFFC with those of the ORFC: the former providing the solid base and the continuity; the latter providing contact with the world at large and a constant influx of new ideas. The process has begun.

Optimism and hope are not the same thing. Right now it is hard to be optimistic but we must never give up hope. The agrarian renaissance is still possible and if it operates as it must at all levels - including moral, economic and metaphysical, as outlined in Chapter 4 - then it truly would transform the way the world is run and its prospects for the future.

But the people to whom we have ceded the most power are not going to bring about the necessary changes. They are wedded to the world as it is. It's up to us, people at large, to do what needs doing. We certainly have the ability. It was the collective genius of humanity that created agriculture in the first place - fashioned crops out of wild plants that were generally fibrous and often toxic, and livestock from wild beasts that were always skittish and often dangerous; and then transformed what those crops and livestock provided, with a little help from still-wild creatures, herbs and shellfish and all the rest, into the world's great cuisines. All we need now is a little more self-belief, to take back what is rightfully ours and build the world afresh.

# Resources

## Publications

The following includes full details of publications cited, chapter by chapter, plus others that I have found valuable over the years:

### 1: What is and what could be

McCance, R. and Widdowson, E. (2002). *The Composition of Foods: sixth summary edition*. Royal Society of Chemistry: London. First published 1940.

Evans, G. E. (1956). *Ask the Fellows Who Cut the Hay*. Faber & Faber: London.

### 2: The future belongs to the gourmet

Tudge, C. (1980). *Future Cook*. Mitchell Beazley: London. (Published in the US by Crown, New York, as *Future Food*). The only book I know that tries explicitly to show the direct relationship between food farming, sound nutrition and great cooking.

### 3: Why don't we do the things that need doing?

George, S. (1976). *How the Other Half Dies: The real reasons for world hunger*. Penguin: London. Reprinted 1986, 1991.

Shiva, V. (2014). *The Vandana Shiva Reader*. University of Kentucky: Kentucky. Brings together essays from one of the world's deepest thinkers in food and agriculture, a former quantum physicist turned fierce campaigner who among other endeavours founded Navdanya (a network of seed keepers and organic producers spread across India), dedicated to human wellbeing and biodiversity.

Lawrence, F. (2004). *Not on the Label: What really goes into the food on your plate*. Penguin: London.

Lang T., Heasman M. (2004). *Food Wars: The global battle for mouths, minds and markets*. Earthscan: London.

Tansey G. and Rajotte T. (2008). *The Future Control of Food: A guide to international negotiations and rules on intellectual property, biodiversity and food security.* Earthscan: London.

Druker, S. M. (2015). *Altered Genes, Twisted Truth: How the venture to genetically engineer our food has subverted science, corrupted government, and systematically deceived the public.* Clear River Press: Salt Lake City. Here is a devastating exposé of GMOs and the whole biotech industry and the governments who support it, showing just how specious the whole exercise has been - and how science, seen as a stronghold of intellectual probity, can in effect be bought.

Sen, A. (1962). 'An Aspect of Indian Agriculture'. *Economic Weekly* 14.

Ünal, F. G. (2006). 'Small Is Beautiful: Evidence of Inverse Size Yield Relationship in Rural Turkey'. *Policy Innovations.* www.policyinnovations.org/ideas/policy_library/data/01382

## 4: Digging deep

Large, M. (2010). *Common Wealth: For a free, equal, mutual and sustainable society.* Hawthorn Press: Stroud. Martin Large discusses the whole concept of the tripartite mixed economy, which surely is what enlightened agriculture really requires.

Elkington, J. (1999). *Cannibals with Forks: The triple bottom line of 21st century business.* Capstone: Oxford.

Ilich, I. (2001). *Tools for Conviviality.* Marion Boyars: London. First published in 1963.

Schumacher, E. F. (1973). *Small is Beautiful: Economics as if people mattered.* Blond & Briggs: London.

## 5: The absolute requirement: fertile soil

Ingham, E. (2004). 'The Soil Foodweb: its role in ecosystem health' in Elevitch, C. R. (ed) *The Overstory Book: Cultivating connections with trees (2nd edition).* Permanent Agriculture Resources: Holualoa.

Falk, B. (2013). *The Resilient Farm and Homestead: An innovative permaculture and whole systems design approach.* Chelsea Green Publishing: Vermont.

Nelson, M. (2014). *The Wastewater Gardener: Preserving the planet one flush at a time.* Synergetic Press: Santa Fe.

### 6: The staples: arable

Briggs, S. (2008). *Organic Cereal and Pulse Production: A complete guide.* Crowood Press: Marlborough.

Diamond, J. (1997). *Guns, Germs, and Steel: A short history of everybody for the last 13,000 years.* W W Norton: New York.

### 7: Livestock I: the basics

Lymbery, P. and Oakeshott, I. (2014). *Farmageddon: The true cost of cheap meat.* Bloomsbury: London. On the general state of world agriculture with particular reference to animal welfare.

Harvey, G. (2008). *The Carbon Fields.* Grass Roots Press: Edmonton. He introduces the idea of mob grazing - and see his other books.

### 8: Livestock II: change for the better

Harvey, G. (ibid).

Savory, A. and Butterfield, J. and Bingham, S. (2006). *Holistic Management Handbook: Healthy land, healthy profits.* (Island Press: Washington). Savory spells out the entire philosophy and practice of grassland management - and see his other books.

Professor Wood-Gush, D. G. M. and Monaghan, P. (1990). *Managing the Behaviour of Animals.* Chapman and Hall: London.

Wemelsfelder, F. (2001). She provides a fine introduction to her ideas in 'The inside and outside aspects of consciousness: complementary approaches to the study of animal emotion'. *Animal Welfare* 10: 129-39.

Schumacher, E. F. (ibid).

### 9: Horticulture

Mollison, B. and Slay, R. M. (1997). *Introduction to Permaculture.* Tagari Publications: Sisters Creek. And see other books by Mollison.

Laughton, R. (2008). *Surviving and Thriving on the Land: How to use your time and energy to run a successful smallholding.* Green Books: Cambridge.

Fukuoka, M. (2009). *The One-straw Revolution*. New York Review of Books: New York. First published in 1975.

### 10: Agroforestry

Gordon, A. M. and Newman, S. M. (1997). *Temperate Agroforestry Systems*. Cabi: Wallingford.

### 11: The Absolute importance of food culture

There are many excellent food writers including some who get deep into the cultures that have given rise to the world's cuisines, such as Claudia Roden, Elizabeth David and Jane Grigson. The following three histories start with the culture and work outwards:

Hartley, D. (1954). *Food in England.* Macdonald and Janes: London.

Wilson, B. (2012). *Consider the Fork: A history of how we cook and eat.* Basic Books: New York.

Burnett, J. (1966). *Plenty and Want: A social history of diet in England from 1815 to the present day.* Thomas Nelson: London.

### 12: Six steps back to land

Seymour, J. (1961). *The Fat of the Land.* Faber & Faber: London.

Salatin, J. (1998). *You Can Farm: The entrepreneur's guide to start and succeed in a farming enterprise.* Chelsea Green Publishing: Burlington. See his other books too.

### 13: Land, money and the law

McIntosh, A. (2004). *Soul and Soil: People versus corporate power.* Aurum Press: London.

Fresh Start Land Enterprise Centre. (2015). *Land Partnerships Handbook. www.freshstartlandenterprise.org.uk/wp-content/uploads/2015/07/LP-Handbook-2nd-Edition-Final-Print-Web-Version.pdf*

*The Land* is not a book but 'an occasional magazine about land rights' which emerges from Bridport, Dorset, several times a year and is essential reading for all who take agrarian renaissance seriously. *www.thelandmagazine.org.uk*

## Reports

The following reports are also highly pertinent:

*Agriculture at a Crossroads: Global report.* (2009). Island Press: Washington. www.islandpress.org/book/agriculture-at-a-crossroads Produced in 2008 by an international body known as IAASTD, co-chaired by Hans Herren and sponsored by the FAO, GEF, UNDP, UNEP, UNESCO, The World Bank and WHO. *Agriculture at a Crossroads* is as authoritative as it is possible to be (participants included experts from 110 countries) and it emphasizes traditional farms and farming for providing the world's food and the need to support and build upon them: farms that in general are small, mixed, low input and skills-intensive, as advocated in this book.

*Foresight. The Future of Food and Farming* (2011). Final project report. The Government Office for Science: London. www.gov.uk/government/uploads/system/uploads/attachment_data/file/288329/11-546-future-of-food-and-farming-report.pdf The *Foresight* report claimed to be building on the IAASTD report of 2008 but reversed its message. It acknowledged that traditional farms have a role but advocated high-tech large-scale industrial farming. The government's chief scientific adviser Sir John Beddington introduced the report with a warning that we cannot continue with 'business as usual' - but then, with a few concessions here and there, recommended that we do just that. *The Future of Food and Farming* has become the British government's (Defra's) standard reference. It says what the government wanted to hear.

*Feeding the Future* (2014). The Landworkers' Alliance. www.landworkersalliance.org.uk/wp-content/uploads/2013/01/20150421-Feeding-the-Future-Landworkers-Alliance-A4-web.pdf

Ponisio, L. C., M'Gonigle, L. K., Mace, K. C., Palomino, J., de Valpine, P., Kremen, C. (2014). 'Diversification practices reduce organic to conventional yield gap'. *Proceedings B.* Royal Society Publishing. rspb.royalsocietypublishing.org/content/282/1799/20141396

Devlin, S. and Dosch, T. and Esteban, A. and Carpenter, G. (2014). *Urgent Recall: Our food system under review.* New Economics Foundation: London. www.neweconomics.org/publications/entry/urgent-recall

Conaty, P. and Large, M. (eds) (2014). *Commons Sense: Co-operative place making and the capturing of land values for 21st century garden cities.* Co-operatives UK: Manchester. www.archive.org/details/Commons_Sense-Co-operative_place_making_and_the_capturing_of_land_value_for_21st

*Financing Community Food: securing money to help community enterprises to grow* (2013). Sustain: London. www.sustainweb.org/publications/financing_community_food

## Essential websites

The following is a shortlist of websites from among the scores that are helpful, of initiatives that my wife, Ruth, and I either founded or helped to found, plus other organizations with which we have close dealings.

The Campaign for Real Farming: **www.campaignforrealfarming.org**
My wife, Ruth, and I started the Campaign for Real Farming website in 2008 with generous help from friends. The campaign has given rise to or fed into all our other initiatives: the Oxford Real Farming Conference, Funding Enlightened Agriculture and (not yet up and running) the College for Enlightened Agriculture and Food Culture. The website includes my own blog, Colin's Corner.

FEA network: **www.feanetwork.org**
See in particular the resource list which covers governance and legal structures, funding mechanisms and community investment, community food projects and community-supported agriculture, access to land.

The Oxford Real Farming Conference: **www.orfc.org.uk**
Includes summaries of key presentations and videos from past conferences, plus news of the next conference.

The Landworkers' Alliance: **www.landworkersalliance.org.uk**

The Pasture-Fed Livestock Association: **www.pastureforlife.org**

Community Supported Agriculture:
**www.communitysupportedagriculture.org.uk**

Shared Assets: **www.sharedassets.org.uk**

Ecological Land Co-operative: **www.ecologicalland.coop**

Soil Association: **www.soilassociation.org**

The Permaculture Association: **www.permaculture.org.uk**

Sustain - The Alliance for Better Food and Farming: **www.sustainweb.org**

Plunkett Foundation: **www.plunkett.co.uk**

Federation of City Farms and Community Gardens: **www.farmgarden.org.uk**

The Biodynamic Land Trust: **www.biodynamiclandtrust.org.uk**

Centre for Agroecology, Water and Resilience: **www.coventry.ac.uk/research/areas-of-research/agroecology-water-resilience**

Schumacher College: **www.schumachercollege.org.uk**

Food Systems Academy: **www.foodsystemsacademy.org.uk**

Compassion in World Farming: **www.ciwf.org.uk**

World Animal Protection: **www.worldanimalprotection.org.uk**

Gaeia: **www.gaeia.com**

Ethex: **www.ethex.org.uk**

Triodos: **www.triodos.co.uk**

Buzzbnk: **www.buzzbnk.org**

## Featured farms

Hornton Grounds Farm: **www.horntongrounds.co.uk/farming**

Maple Field Milk: **www.facebook.com/pages/Maple-Field-Milk/324571964385153**

Chagford: **www.chagfood.org.uk**

Organiclea: **www.organiclea.org.uk**

Worton Organic Garden: **www.wortonorganicgarden.com**

Stream Farm: **www.streamfarm.co.uk**

Tamar Grow Local: **www.tamargrowlocal.org**

Growing Communities: **www.growingcommunities.org**

# Index

# Also by Green Books

## Farm Digesters:

Anaerobic digesters produce clean renewable biogas, and reduce greenhouse emissions, water pollution and dependence on artificial fertilizers

*Jonathan Letcher*

Farm digesters are best known as a means for producing biogas. However, anaerobic digesters are much more beneficial, because they represent a significant way to make both our energy supplies and our food production more sustainable whilst enabling us to run cost-effective farms.

Accessibly written, this book is a must-read for anyone who is running a farm, cares about creating a sustainable future, or is involved in planning for agriculture.

## Surviving and Thriving on the Land:

How to Use Your Spare Time and Energy to Run a Successful Smallholding

*Rebecca Laughton*

This book looks at ways in which projects can be designed that care for the people involved in them as well as the earth that they are trying to protect. If land-based ecological projects are to offer a realistic solution to the problems we face in the twenty-first century, it is imperative that they should be sustainable in terms of human energy. This book offers a framework, backed up by real life examples, of issues to consider when setting up a new project, or for overcoming human-energy-based problems in existing projects.

## About Green Books

Environmental publishers for 25 years. For our full range of titles and to order direct from our website, see: **www.greenbooks.co.uk**

Send us a book proposal on eco-building, science, gardening, etc.
- **www.greenbooks.co.uk/for-authors**

For bulk orders (50+ copies) we offer discount terms. Contact **sales@ greenbooks.co.uk** for details.

Join our mailing list for new titles, special offers, reviews and author events: **www.greenbooks.co.uk/subscribe**

 @ Green_Books  /GreenBooks